Kepler's Physical Astronomy

Bruce Stephenson

Kepler's Physical Astronomy

Princeton University Press
Princeton, New Jersey

Published by Princeton University Press, 41 William Street,
    Princeton, New Jersey 08540
In the United Kingdom: Princeton University Press, Chichester,
    West Sussex

Library of Congress Cataloging-in-Publication Data

Stephenson, Bruce.
    Kepler's physical astronomy / Bruce Stephenson.
        p.   cm.
    Includes bibliographical references and index.
    ISBN 0-691-03652-7 (pbk.)
        1. Astronomy.   2. Kepler, Johannes, 1571–1630.   I. Title.
    QB43.2.S74   1994
    520—dc20   93-38820

ISBN 0-691-03652-7

First Princeton Paperback printing, 1994

Printed in the United States of America

Preface

This book began more than 10 years ago as a modest attempt to understand some of the technical details of the physical theories in Kepler's *Epitome*. As I expanded the study into a doctoral thesis, and particularly as I grappled with the *Astronomia nova*, I found that I could not explain the physics to my satisfaction without adopting a more comprehensive attitude, and treating the *Astronomia nova* as a single extended argument. Kepler believed his argument to be ultimately about the design of Creation; but in the *Astronomia nova*, he presented it as physical astronomy—a new science.

In these pages, I try to explain how the physical astronomy works, for surprisingly little of it has been explained before. I also try to show how Kepler's detailed investigations were always directed at the larger goal of understanding the true design of the heavens. I hope that readers who have hitherto regarded Kepler as a diligent condenser of data, or as a numerically inclined mystic, will learn here to appreciate the intelligence and integrity with which he pursued scientific truth.

Noel Swerdlow awakened my interest in the history of science over a decade ago. He has taught me everything I have learned about the history of the exact sciences. During the years I struggled to understand Kepler, he spent uncounted hours with me reading and discussing my rough drafts. I can see the influence of Swerdlow's understanding of ancient astronomy on every page of this book. Its completion is due in no small part to his intellectual support and to the emotional support implicit therein.

I also thank my friends who encouraged me as I worked for years on what most of them must have thought a terribly obscure subject. I will not name them all here. They know who they are and I hope they know how much they helped. My parents, Charles V. and Luellen H. Stephenson, have consistently amazed me with their ability to encourage without applying pressure. I dedicate this book to them and to my wife Marija.

Bruce Stephenson

Contents

Chapter 1

Introduction

Astronomy was the first natural science, by a margin of nearly two thousand years, to be developed into the form of precise mathematical theory. Ancient astronomy—by which I mean, for present purposes, Greek astronomy and its heritage in the Arabic and Latin West—used geometrical models and their numerical parameters, derived with painstaking trigonometry, to predict the movements of the sun, moon, and planets. This science culminated in the work of Johannes Kepler (1571–1630), Imperial Mathematician at Prague, and discoverer of three results so important that they are known as the laws of planetary motion. Kepler essentially solved the problems of ancient astronomy: not only were his models much more accurate than their predecessors, they were innovative theoretically to such an extent that they stand quite alone. Half a century after Kepler died, Isaac Newton swept away the older science by showing how to derive astronomical results from his more general dynamics. Thus Kepler was the last great practitioner of the science of ancient astronomy.

In addition to treating so decisively the problems of the old astronomy, Kepler played a major part in the creation of the modern science of astronomy; and this in two ways. First, his so-called "three laws," once established, provided the solid empirical foundation for Newton's great *Principia* (1687), the book with which modern astronomy begins. Newton was able to derive, from his own mathematical and physical principles, these three most remarkable of Kepler's discoveries. Planetary orbits were ellipses; the time required to traverse any arc of an orbit was proportional to the area of the sector between the sun and that arc; and the cubes of all the planets' mean distances from the sun were proportional to the squares of their respective periodic times. By deriving these three results, and by presenting them as evidence for his own theories, Newton rescued them from the oblivion of a science superseded, even as he was establishing beyond all doubt that the successor to ancient astronomy would be a part of physics.

The other sense in which Kepler's work led toward modern astronomy, even as it completed the old, is that he was the first actually to envision astronomy as a part of physics. As we shall see, Kepler was the first astronomer forced to confront planetary motion as a physical problem. An essential element of his theories was the physics underlying them. This physics was flawed at the foundations, for he lacked the modern concept of inertia (in the slow development of which his contemporary Galileo was then engaged). Yet the physical theories, if not fruitful for physics proper, proved a reliable guide in his own astronomical researches. Kepler's mathematical ingenuity was such that he could invent alternative theories, each good enough to "save" even the excellent observations of Tycho Brahe. These observations, therefore, did not provide enough guidance for him to reach his final solution to the problem of planetary motion. He was quite capable, as we shall see, of inventing other theories which were observationally adequate. The necessary guidance he took from physics, rejecting theories for which he could find no plausible physical basis, and doggedly analyzing the interplay of forces where he could. Had he not devoted himself tirelessly to physics, he would certainly not have discovered the laws of planetary motion; but more than that, he would not even have asked the questions (for they were new questions) to which these "laws" were his answer.

I have just now made a large assertion. The present study, in attempting to justify it, will constitute a major new look at Kepler's astronomy. It will yet be incomplete, from Kepler's own point of view, for I shall not touch deeply upon the archetypes underlying his universe, nor on the harmonies expressed in it, and perhaps for him these subjects were the most important in all of astronomy. Neither shall I examine his extraordinary technical skill at bringing the raw observations to use in the development of astronomical theory, an area where no one since Ptolemy approached him in originality. Instead, I shall concern myself with the area where Kepler's chief importance for the history of science lies: the development of his planetary theory.

I will attempt to explore and explain the development of Kepler's planetary theory, and of the physical hypotheses integral to that theory, more faithfully than has yet been done. My account will be faithful to Kepler in taking seriously his "incorrect" theories along with the "correct" ones, unlike the early, technically accomplished studies by Delambre, Small, and others; and in taking seriously the technical depth of these theories, unlike most recent studies. The largest part of this account, of course, will be a detailed examination of the great *Astronomia nova* of 1609.[1] This profoundly original work has been portrayed as a straightforward account of converging approximations,

---

[1] Volume 3 in *Johannes Kepler Gesammelte Werke*, ed. W. von Dyck, M. Caspar, and F. Hammer (Munich: C. H. Beck, 1937–). I will normally cite Kepler's works in this edition, with the abbreviation *G. W.*, giving volume, page number, and where appropriate line numbers.

and it has been portrayed as an account of gropings in the dark. Because of the book's almost confessional style, recounting failures and false trails along with successes, it has in most cases been accepted as a straightforward record of Kepler's work. It is none of these things. The book was written and (I shall argue) rewritten carefully, to persuade a very select audience of trained astronomers that all the planetary theory they knew was wrong, and that Kepler's new theory was right. The whole of the *Astronomia nova* is one sustained argument, and I shall make what I believe is the first attempt to trace that argument in detail.

Another major focus of this study will be Kepler's lunar theory. Kepler developed an ingenious account of physical processes that would explain the four known inequalities in lunar observations, and elaborated it with rather elegant geometrical constructions to calculate lunar positions from the physics. These theories are unknown in the historical literature on Kepler. I shall take care to give them, at last, the attention they deserve.

More generally, I will attempt throughout to show that many of Kepler's odder hypotheses, such as those concerning possible celestial "minds" governing planetary motion, far from being gratuitous speculation, played an important role in his physical analysis and in the argument of the *Astronomia nova*. Whatever he may have thought about the actual existence and nature of such entities, Kepler used them hypothetically to analyze problems in abstract physical terms. When he argued that a mind in a planet could not direct itself (or that planet) on an eccentric circular orbit, he was also saying that, *a fortiori*, no plausible physics could produce that pattern of motion. I choose to discuss such interpretations, which bring out the use of a passage in Kepler's larger argument, rather than digress pointlessly about spiritual beings, on the one hand, or cover up the whole matter, on the other.

I have already stated some of the areas in which this study is incomplete. There is another, larger sense in which this is so. I do not compare Kepler's work with that of contemporary astronomers or natural philosophers, nor do I discuss their reactions to it. My account, therefore, may seem to lack context. This is a serious accusation to any historian, and one on which I feel obliged to comment. In studying any writer, one should normally examine the context in which he or she worked, in order to clarify the problems being addressed, and also the accepted meaning of the terms used in the discussion. For these purposes, however, the context most relevant to Kepler was not his contemporaries, but rather the small and sophisticated world of ancient astronomy: Ptolemy, Peurbach, Regiomontanus, Copernicus, and Maestlin, to name only those most important for Kepler. This scientific field had an established and well-developed technical vocabulary and set of problems and techniques. I do, in fact, inevitably discuss this context, although I do not attempt a general history.

Regarding "physics" the situation was otherwise. The physical theory available to Kepler was very much cruder than the astronomy, less definite and less likely to be remembered today. Wider research would no doubt reveal significant nuances in the meaning of his physical vocabulary. My subject is already a large one, however, and it is not physics but the use to which he put physics in his astronomy. Here Kepler went his own way. One has only to read the letters he exchanged with David Fabricius, who was practically his only serious correspondent on astronomical matters during the years of the *Astronomia nova*, to see how lonely Kepler was in his work. Fabricius—and much more his other contemporaries—simply did not understand what Kepler was trying to do. Even if we are not certain exactly how he was bending his physical vocabulary to his needs, we can follow his attack upon the essentially new problems of developing physical theory to accompany his developing astronomy. The astronomy itself would be almost unintelligible, in its complexity, without the context of Ptolemy and Copernicus. The use of physical analysis was newly invented by Kepler, and that is how we must see it.

The Cohesion of the Earth

Kepler's best-known essay at physical theory is his explanation, in the introduction to the *Astronomia nova*, of why the earth and all the various objects on it hold together. In Aristotelian physics this had been no problem: heavy things moved of their own accord to the center of the universe, and the center of the universe was inside the earth. This easy explanation was no longer available to Kepler, as a Copernican. He therefore sketched, in his introduction, a theory of heaviness (*doctrina de gravitate*) which was consistent with the earth moving and being elsewhere than at the center. His principles were simple. First, objects did not move of their own accord, but instead tended to stay wherever they were, so long as they were outside the sphere of influence of a "like body." Secondly, heaviness was a disposition of like bodies (*cognata corpora*) to unite, each being attracted in proportion to the other's mass.

This sounds like the makings of a theory of gravitation, and has been hailed as such by many scholars, including no less a figure than Delambre. Kepler posited an attractive virtue or power (*virtus tractoria*) that was mutual and proportional to mass; and he further concluded that the moon's attractive virtue was responsible for the tides. His definition of lightness as a merely relative lack of heaviness is not entirely congruent to modern ideas; but it yields a correct explanation of why light things rise, and after all a tendency to rise was the common characteristic denoted in the old physics by the word lightness, *levitas*. To be sure, the theory had problems. Kepler explicitly supposed the attractive virtue to be of finite range, and he limited its effect to like bodies, without explaining just what he meant by like bodies. He hinted that the force of attraction weakened with distance, but did not elaborate. (Elsewhere he did remark that *light* varied as the inverse square of distance, thereby tantalizing some later historians but hardly anticipating the law of gravitation.)

My discussion of Kepler's physical astronomy will not encompass this so-called theory of gravitation. Kepler has been described[1] as straddling the watershed between medieval and modern science, a characterization which applies readily to his physical investigations. Physics in medieval Europe, as derived primarily from Aristotle, was for the most part a search for causes. The new physics created in the seventeenth century, primarily by Galileo and Newton, turned to the question, less profound perhaps but certainly more fruitful, of finding the unifying principles from which observed phenomena could be mathematically deduced. Kepler devoted large amounts of time to both kinds of physics, and applied the word "physical" indifferently to either. I shall have relatively little to say about his quest for causes, concentrating instead upon the insight into astronomy he gained by postulating—for the first time, really—physical principles behind the motions in the heavens; and upon his deduction of astronomy from those principles.

The physical principles themselves, let it be said at the outset, were wrong in virtually all of their particulars. The astronomy was startlingly good, far the best that had ever been created. We should not marvel, then, at the difficulties he had in deducing one from the other, particularly since the mathematics available to him could be applied to his problems only in the most tedious and circuitous ways. That he achieved such considerable success in relating phenomena to physics is no small tribute to his industry and intelligence.

As to his *doctrina de gravitate*, it is better thought of as an episode in the struggle for heliocentrism than as a step toward universal gravitation. About one-third of the introduction to the *Astronomia nova* actually introduced that book; the remainder advocated, and defended, the Copernican system of the world. The arguments in favor of heliocentrism urged its simplicity, elegance, and physical plausibility. Most of them depended upon some knowledge of astronomy, and several upon the results of Kepler's own investigations. They were good and subtle arguments, and we shall encounter many of them in the following pages. Against Copernicus the arguments were comparatively simple, but two of them in particular, if we imagine ourselves at the turn of the seventeenth century, were undeniably potent. The first was the problem of how the earth held itself together and held heavy objects to it; the second was the objection from scripture, which seemed to state clearly that the sun moved around the earth. Kepler in the introduction devoted rather more attention to refuting the scriptural objections, but did write several pages on the earth's coherence. These are the pages containing the *doctrina de gravitate et levitate*.

Now, the coherence of the earth and its affixed objects was no small problem, once one had given up the idea that heavy objects fall to the center of the world because that is their natural place. No satisfying explanation of the phenomena we attribute to gravitation was found until Einstein's general theory of relativity, which reduced them to geometry. Prior to that theory

---

[1] A. Koestler, *The Sleepwalkers* (London: Hutchinson, 1959).

there was little alternative, if a cause was required, to the bald assertion that things simply attract one another. It is not, of course, for any such trivial remark that we ascribe the theory of gravitation to Newton. Rather it is because Newton applied his extraordinary mathematical talents to the problems of demonstrating that a large class of phenomena could be deduced from a few principles, one of which was a specific and precise formulation of the assertion that things attract one another.

Kepler did, in his defense of heliocentrism, go a couple of steps beyond the claim that certain things attract one another. He suggested that the attraction was proportional to mass, a good guess, not very difficult to think of but certainly a gain in precision. He further specified that the attraction was mutual, a more subtle refinement. Beyond these points he did not go. He thought the attraction to be of finite range and limited to like bodies, perhaps because he thought it similar to magnetism, which William Gilbert had believed to be limited in the same ways. More important, he made no serious attempt to work out the implications of this attractive virtue he had postulated in defense of Copernicus. Although he stated clearly that the range of the earth's attractive virtue extended beyond the moon, he made no use of that attraction in his physical lunar theory. His lunar theory was fashioned after his planetary theory, for the obvious reason that the major components of lunar motion correspond exactly to those of planetary motion. I shall examine both theories in detail, but I shall not have occasion to invoke the theory of the earth's cohesion. It should be superfluous to add that Kepler's knowledge of the inverse-square attenuation of light constitutes in no way an approach to the theory of gravitation. His own discussion of why the solar force attenuated as the simple inverse of distance, rather than the inverse square, tantalizes us only as long as we insist on seeing his theory as a failed attempt to enunciate Newton's theory. When we examine Keplerian physics in its own terms it is clear that his "forces," the components of his theory for which he used the word *vis*, are not analogues of Newtonian "forces." (A Keplerian "force" caused velocity, not acceleration.) Kepler was perfectly correct on this point: his "forces," if we fit them into the mathematics of planetary motion, suffer attenuation only as the direct inverse of distance from the sun. He was, of course, in error at the foundations of his physics. One cannot build a satisfactory physical theory from Keplerian "forces."

Kepler tried to do just that. By shedding the anachronistic presumption that his work ought somehow to line up neatly as a precursor of Newtonian physics, we can learn much from it. The very fact that it is on the wrong track should free us from the temptation to fragment it into discrete "results" which we might compare and contrast with the "correct results." Yet unlike so many misguided theories, Kepler's physics is precise and coherent. We can take it seriously. Since its flaws lay in the foundations—in Kepler's conception of inertia—we can trace, in the detailed development, the creation of a science almost *ex nihilo*. At the same time, if we are careful, we can avoid the distortions which always arise from the complacent, dangerous feeling that we know already what a nascent science is going to turn out to be.

On the other hand, scientific work in the long run is distinctly characterized by progress. In some sense more vexing to philosophers than historians, scientific theories are worked out, and eventually accepted, which are better than the ones they replace. However hard it is to define scientific progress with precision, the most interesting and the most important questions in the history of science pertain to genuinely progressive scientific work. Here is the reason Kepler's physical theories hold so much interest. Not of any lasting value as physical theories, they were yet no fruitless appendage to his brilliant career as an astronomer. They developed concurrently with his astronomy, shaping his view of that science and, in ways that I shall examine, guiding him in his conquest of the problems of ancient astronomy.

The story is an interesting and an important one. More than any other single reason, a misguided and sometimes unwitting interpretation of Kepler's physical astronomy as a precursor to Newton's has impeded any fair analysis of it. I am going to try and analyze Kepler's physical astronomy as the science that it was, fundamentally wrong, but intelligent and original. Without relinquishing the benefits of hindsight, we must avoid confusing Kepler's physics with Newton's. The Keplerian *doctrina de gravitate* was an attempt to provide a cause to replace that which had been lost with geocentrism, and thus to defend the system of Copernicus from a dangerous objection. My topic is instead Kepler's search for physical principles from which, by mathematical demonstration, he could learn the true motions of the planets.

Chapter 2

Mysterium Cosmographicum

In the preface to his first book, the *Mysterium Cosmographicum* of 1596,[1] Kepler summarized the early investigations leading to that book with admirable precision: "above all there were three things of which I diligently sought the reasons why they were so, and not otherwise: the number, size, and motion of the spheres." [2] These were novel questions, and indeed the *Mysterium* was a little book filled with peculiar questions, a book whose purpose was only in part scientific. Kepler poured forth his ideas enthusiastically, and although he was quick to admit their flaws, he was equally quick to excuse them. His exuberance was not yet balanced by the self-criticism which distinguished his mature writings. His technical command of mathematics and astronomy was still insecure. The questions he raised in the *Mysterium*, for all their originality, had no common themes beyond Copernican cosmology and Kepler's desire to understand, through it, the mind of the Creator.

We cannot approach such a book as we will approach Kepler's later, more sophisticated work. A very large part of the *Mysterium* has nothing at all to do with physical astronomy. Accordingly, we shall discuss only the passages where Kepler was working with ideas that were in some sense physical, and not simply mathematical or aesthetic. Even so, we cannot analyze these passages as if they contained sustained argument, for they do not. Kepler was trying out various ways of bringing his physical intuition to numbers, without having decided yet what kind of a theory he was seeking. By examining his attempts, and the numbers he obtained, we can draw a few conclusions, and speculate about the manner in which he arrived at his published hypotheses.

---

[1] The *Mysterium* has been translated into English by A. M. Duncan, with notes by E. J. Aiton (New York: Abaris Books, 1981).
[2] *G. W.*, 1: 9: 33–34.

We can see rather clearly that he was not yet trying to create a full-fledged physical astronomy. Some of his ideas in this book turned out to be quite fruitful in his later career, but here they were not rooted in any overall conception of astronomy.

We point this out now, because later, at the time of the *Astronomia nova*, Kepler quite clearly did conceive of astronomy as a physical science. His purposes in that book were focused to a far greater degree, and accordingly we will be able to give a much more satisfying account of his work. Meanwhile, however, there are mysteries worth probing in the *Mysterium*.

Toward the end of the preface to the *Mysterium*, Kepler described some of his earliest work, preceding the results he reported in the text. He had sought a pattern among the proportions of the different planetary distances from the sun, but had not succeeded, even when he introduced a couple of hypothetical planets, too small to see, in filling out the larger intervals. In retrospect he realized that such methods had an inherent defect: they failed to limit the number of planets. They were not adequately constrained. The space within Mercury's orbit and beyond Saturn's could be further divided, along the same lines, with no apparent end. This was no way to comprehend the heavens.

Kepler's next attempt, considerably more important to us, had been to find a pattern connecting the (linear) speeds[3] of the planets with their distances from the sun. He knew that the outer planets moved more slowly than the inner, but that the speeds decreased more slowly than the distances increased as one moved out from the sun. He guessed that the speed and distance might vary as shown in Figure 1. Here the speed and distance of each planet lie (hypothetically) along a quadrant of a circle. That is, if we introduce the parameter Θ as shown in the figure, he speculated that the ratio of speed to distance was $(1 - \cos\Theta)/(1 - \sin\Theta)$, with a value of Θ particular to each planet (Figure 1). This theory, if we may so term it, had most of the same flaws as the other one (notably a lack of empirical adequacy), but at least it did predict that the fixed stars would be motionless at the (finite) distance corresponding to $\Theta = 0$, as Kepler believed. It further predicted that the force of motion would be maximal, perhaps infinite, at the center where was the sun, representing motion in its very essence.[4]

Although this second theory addressed the physical problem of variation in the planets' motion, and although Kepler phrased it in the explicitly physical terms of a "moving virtue" (*virtus movens*), one cannot detect here any real physical theory. Some sort of moving virtue varied in a mathematically defined way, but Kepler had neither explanation nor analogy for how or why it varied in this way. He was not trying to understand the cause responsible for the different speeds, nor to relate them to any physical process, but only to find a pattern in them. The real physical questions were not

---

[3] Here and throughout the remainder of this study, a planet's "speed" or "velocity" means linear velocity, and not angular velocity, unless specifically noted otherwise.

[4] *G. W.*, 1: 11: 7–21.

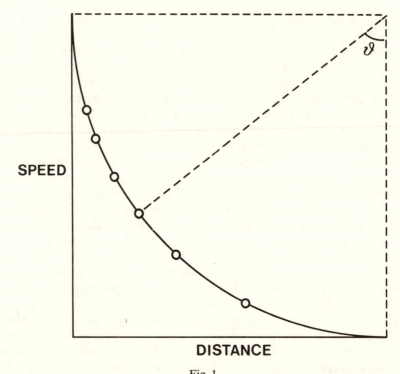

SPEED

DISTANCE

Fig. 1

addressed here in the preface, but only somewhat later, in the closing chapters of the book.

The first half of the *Mysterium* was a speculative examination of the five regular polyhedra and of Kepler's principal thesis that they could be interposed neatly between the spheres of the six planets in the Copernican system, thus determining both the number of planets and the relative sizes of their orbits. (The *Mysterium* was thoroughly and openly Copernican. Indeed, the orbital radii were undetermined in the Ptolemaic astronomy Kepler knew.[5] The Tychonic arrangement, had Kepler known of it at the time, would clearly have been incompatible with his suggested nesting.) After quite a bit of discussion, the abstract considerations of the fitness and harmony of this arrangement gave way abruptly in Chapter 13 to a numerical investigation of the agreement between Copernicus's parameters and Kepler's cosmographic thesis.

---

[5] We now know that Ptolemy did attempt, in his *Planetary Hypotheses*, to determine the orbital radii astronomically (Bernard Goldstein, "The Arabic Version of Ptolemy's *Planetary Hypotheses*," *Transactions of the American Philosophical Society*, new series, 57:4, June, 1967, pp. 3–12). Kepler and his contemporaries were unaware of Ptolemy's physical attempts to determine the orbital radii.

The numbers did not entirely agree, but Kepler was not at all disconcerted. He pointed out that planetary distances, as taken from Copernicus, were computed not from the sun itself but from the mean sun, the center of the great sphere which carried the earth. Using the mean sun as center had the odd consequence that the sphere of the earth had no thickness, unlike all the other planetary spheres, since the earth's path was everywhere equidistant from the mean sun. The dodecahedron outside it and the icosahedron inside it touched the same sphere (or virtually the same sphere; room may have been left, he thought, for the body of the earth itself, and perhaps also for the moon's orbit). No extra space was needed to allow for an eccentricity. This did not seem right. The mean sun was a mathematical construct; surely the planets were arranged around the sun itself. Kepler was not yet able himself to carry out the computations needed to determine the distances from the true sun; but his former teacher Michael Maestlin had done so for him. In the fifteenth chapter, Kepler laid out a table comparing both sets of distances with those derived from the regular polyhedra.

One cannot help noticing in this table that the corrected distances agree no better than the old ones. Still, Kepler used the new distances, without comment, throughout the remainder of the book. The reason, I think, is clear. He was writing not about mathematical hypotheses, but about the real universe, where planets moved in their paths about the sun according to definite rules. Nothing was arbitrary; all was as God intended. The mean sun had indeed been a very important point in the Copernican hypotheses, which were intended in their technical aspects to reproduce the equations of the Ptolemaic models. It had no place in a physical world governed by the sun itself. Kepler later made this argument explicit in the *Astronomia nova*, but it was already present in Chapter 15 of the *Mysterium*, and indeed in the assumptions underlying the whole book. The planets' distances from the mean sun were physically meaningless, and so he simply did not use them.

The regular solids, if Kepler had been correct, would have answered two of his three questions, namely those concerning the number and the sizes of the spheres. Today we no longer expect the number of planets to have an interesting explanation, nor have their orbital radii as such proved a fruitful field of study (although the Titius-Bode law remains a puzzle). Kepler's attempts to understand the third of his problems, the motion of the planets, led him in quite another direction. The regularity he wanted to explain was that the planets, arranged in their Copernican order, were also arranged in order of speed. Planets more distant from the sun moved more slowly, not only in angular velocity—which was to be expected because of their greater distances—but absolutely.

Kepler approached this question as supplementary evidence for the heliocentric arrangement, as if he were not certain what business it had in a book on the cosmographic mysteries of the Platonic solids. In the early part of the book he had alluded several times to the support which his theory of the regular solids gave to the Copernican hypothesis. After developing that theory as far as he could, he opened Chapter 20 with the remark that there was perhaps another argument, taken from the motion of the planets, by which the new hypothesis could be confirmed. We cannot doubt his sincerity as an advocate for Copernicus, but there was more motivation than this behind the three ensuing chapters on the weakening of planetary motion with distance from the sun. Kepler was trying to understand the creation of the heavens, and the motions in the heavens were part of the plan by which they had been created.

More distant planets took longer to complete their revolutions. The periodic times, however, increased not simply in proportion to the size of the orbit, as they would if all the planets moved at the same speed, but yet more, indicating that the distant planets were slower in absolute speed. Always the large intervals in distance corresponded to the large differences in time, and the small to the small. Kepler concluded that the planets' speeds were dependent on the distances. Either the motive souls (*motrices animae*) which were more remote from the sun were also weaker, or there was but a single motive soul in the sun whose ability to move the planets was attenuated with distance. As to what sort of thing this motive soul might be, to which he was attributing the revolutions of the planets, Kepler was silent. He referred to the sun's virtue (*virtus*); and he stated that the sun greatly exceeded the other parts of the world in the efficacy of its virtue (*virtutis efficacia*); and he made several other remarks trying to show why it was reasonable that distance from the sun mattered (and why, not incidentally, one should take the sun itself as the center). He did not try to explain how the sun, or a motive soul in it, could move distant planetary bodies. Instead he took up the mathematical relation between distance and periodic time, attempting at least to ascertain how the soul, or whatever, worked at different distances.

He was not able to say a lot about this question. His efforts to connect the distances and times were clumsy and particularistic, in the same sense that his use of the five perfect solids had been particularistic. His inability was perhaps of no great consequence. Kepler was, if not really on the path which would lead to his "third law," at least looking for that path; but as anyone who has tried can testify, he could not have fit all of the data in Chapter 20 of the *Mysterium* to a 3/2-power law. Some of the distances he used (those for Saturn and Mercury) were simply wrong. He had asked a surprisingly good question, but was neither mature enough nor adequately supplied with data to answer it.

Kepler's raw data, as he presented them, are reproduced in Table 1.[6] Each number is the time, in days, which would be required to traverse the orbit indicated at the left, by traveling at the velocity of the planet at the top. Thus a planet moving at the velocity of Saturn would require 1174 days to complete the earth's orbit; at the velocity of Jupiter, 843 days to do the same; at the

---

[6] *G. W.*, 1: 69. The number 1282, for Jupiter's velocity in the orbit of Mars, is evidently a computational error and should be 1256.

Table 1. Hypothetical periodic times

		Velocity					
		Saturn	Jupiter	Mars	Earth	Venus	Mercury
	Saturn	10759					
	Jupiter	6159	4333				
Orbit	Mars	1785	1282	687			
	Earth	1174	843	452	365		
	Venus	844	606	325	262	225	
	Mercury	434	312	167	135	115	88

velocity of Mars, 452 days; and so on. The outer planets, obviously, are slower. The table is computed from the observed periodic times, which appear on the diagonal; and from the Copernican ratios of distances from the sun, which determine the proportions down each column. These latter are simply not good enough for Kepler to have drawn any reliable quantitative conclusions.

Nevertheless, Kepler's attempts are interesting for the light they shed upon his early attitude toward scientific explanation. The distance of a planet from the sun evidently was related to its periodic time in two ways, first because more distant orbits are simply larger, and secondly for the physical reason that more distant planets move slower. "Therefore it follows," he concluded, "that a single greater distancing of the planet from the sun works twice toward increasing the period, and on the other hand, the increment of the period is double (*duplum esse ad*) the difference of the radii."[7] Kepler wanted to argue, one would think, that the orbital length increased directly with distance, the speed varied inversely with distance, and the time, their quotient, therefore varied as the square of the orbital radius. Should we not have translated "*duplum*" as "the square of" rather than "double"? This would have made sense, but was not in fact what Kepler meant. He clarified, "Thus half the increment added to the lesser period should give the true proportion of the distances."[8]

A worked example and a table of predicted distances leave no doubt. If we represent the periodic times of a planet and the one immediately superior to it by T_i and T_s, and their distances from the sun by R_i and R_s, Kepler believed it likely that

$$\frac{R_s}{R_i} = \frac{T_i + \dfrac{(T_s - T_i)}{2}}{T_i} = \frac{\dfrac{T_i + T_s}{2}}{T_i} \tag{1}$$

[7] *G. W.*, 1: 71: 13–16.
[8] *G. W.*, 1: 71: 17–18. Kepler always preferred to formulate his mathematics in terms of the time elapsed in a portion of the orbit (the "delay") rather than the velocity, as discussed later, and in such formulations, halving the increment to the period comes naturally.

Table 2. Ratios of adjacent planetary distances

Radius of		Calculated	Observed
Jupiter,	as proportion of Saturn	.574	.572
Mars,	as proportion of Jupiter	.274	.290
Earth,	as proportion of Mars	.694	.658
Venus,	as proportion of Earth	.762	.719
Mercury,	as proportion of Venus	.563	.500

The proportional increase in radius was arithmetically half of the proportional increase in periodic time. Kepler's computations gave the distances as shown in Table 2.[9] With the given numbers, the agreement was close enough to provide some support for the theory. (In fact, the first and last of the "observed" ratios are not accurate.)

Considered as a physical deduction, though, equation (1) is most peculiar. We can only speculate where Kepler got the idea of "halving" a proportional increase by taking half the increment, instead of the square root of the proportion. If he had reasoned that periodic time was as the square of distance, he would have reached a much more plausible relation:

$$\frac{R_s}{R_i} = \frac{\sqrt{(T_i \times T_s)}}{T_i} = \sqrt{\frac{T_s}{T_i}} \tag{2}$$

However, plausible as equation (2) may be, computations based on it diverge from the observed sizes of the orbits much more than do those based on equation (1), and so Kepler had every reason to refrain from making the "sensible" interpretation of his physical insight. Many of the logical weak points in Kepler's physics occur where observational facts seem to have forced him into positions he might not otherwise have taken.

The oddest thing, to us, about relation (1) as Kepler stated and used it is that it totally lacks the character of a *general* law. One can compute the radius of Venus's orbit compared to Mercury's from their periodic times, and likewise the radius of the earth's orbit compared to that of Venus; but computing the radius of the earth's orbit directly from that of Mercury would not give the same answer. This is, of course, a consequence of the mixing of geometric and arithmetic progressions in (1). If Kepler's relation was an accurate description of the orbital sizes and periods, it was so only for the existing configuration of orbits. It was not, and could not be, a *general* law. From the correct relation we know as Kepler's "third law" (or, for that matter, from the mathematically sensible equation [2]), one can show that a planet's velocity is some function of its distance from the sun. It is then meaningful to consider the speed of a hypothetical planet at an arbitrary distance from the sun. Kepler's equation

---

[9] *G. W.*, 1: 71. The "observed" ratios are consistent with the ratios of the means of aphelial and perihelial distances in the second column of Kepler's table in Chapter 15, *G. W.*, 1: 54.

(1), on the other hand, cannot be made to yield any such simple relation. Any expression you derive for the velocity of a planet from (1) will contain a parameter from the orbit of another planet. The supposition of a hypothetical or arbitrary planet would disrupt the whole system.[10]

I have no evidence Kepler knew that his theory of Chapter 20 was so intimately connected to the particular configuration of the solar system. I am inclined to think he did; if for no other reason, simply from trying out the relation (1) on various configurations of planets. Certainly a great deal of exploratory calculation lay behind the *Mysterium*. It is worth noting, in this regard, that equation (1) distinguishes between the superior and inferior of two adjacent planets. If you interchange the subscripts in (1), and compute the size of Mercury's orbit from the two periodic times and the orbit of Venus, the orbits will be in a different proportion. Kepler nowhere mentioned this; and he might have found it awkward to explain why one had to work from the inside out and not the other way around. But he carefully distinguished the inner and outer planets in his statement of the relation; and if you do it backwards, working from the outside in, the radii you get are *much* worse, hopelessly far from the Copernican values.

If Kepler was doing a lot of calculation, guided by both his physical intuition and his Copernican parameters, trying to relate the distances and times, then he could well have tried, and rejected for empirical reasons, the seemingly natural formulation of his insights as equation (2); and he could also have tried the inward- and outward-moving versions of the more forced formulation (1), rejecting the former but keeping the latter as a reasonably good match to the data. This is an attractive possibility, as it would relieve our dismay at Kepler's use of half a proportional increment, where the square root of the proportion seems called for by his own logic; but we shall probably never know.

Any numerical testing of this type of relation would have revealed what we have already remarked, that if relation (1) holds for a particular set of planets, it would not have held if some of them were removed, nor if new planets were introduced among them. We know from the introduction to the book that Kepler had at one time experimented with configurations involving extra planets, while he was trying to answer that other question about the sizes of the orbits. He had been pleased, however, when his solution to that question, namely the interposed polyhedra, had not only dispensed with unknown planets, but had even fixed the number of possible planets at six. It is safe to assume that Kepler was probably pleased when he found that the theory of Chapter 20 could tolerate no new or arbitrary planets among the known ones. Certainly at the time he wrote the *Mysterium* he was not searching for abstract physical laws, but was rather trying to explain the reasons by which God created the universe as it was, with a sun and six planets moving around it,

---

[10] As a matter of fact, Kepler's "third law" also shares this characteristic, in the context of his physical astronomy. It was not a relation that *he* could apply to hypothetical or arbitrary planets.

along specific orbits at specific velocities. He was trying to solve the particular problem of our universe, not the general one of possible universes.

The reasoning by which Kepler arrived at the table corresponding to equation (1) was based on periodic times and solar distances; physical insight had entered only with the assertion that distance played a double role in determining the period. Toward the end of the same Chapter 20 he attempted to work directly with the forces causing motion. He hypothesized that the virtue moving the planets was attenuated in the same way that light would be as it spread outward in circles. That is, the strength of the motive virtue was in inverse proportion to distance from the sun. Kepler took it as probable that "any planet, however much it exceeds a superior in strength of motion (*fortitudo motus*), is exceeded by that much in distance."[11] Now, almost as if to give equal time to the individual-motive-souls theory of weakened motion, he phrased an example in terms of the Martian virtue (*virtus Martis, virtus Martia*) or force (*vis Martia*), as compared to that of the earth. We would today express his assumption, that the ratio of the distances is that of the speeds inversely, by writing an equation such as (3):

$$\frac{R_m}{R_e} = \frac{V_e}{V_m} = \frac{R_e/T_e}{R_m/T_m} = \frac{R_e \times T_m}{R_m \times T_e} \tag{3}$$

Kepler's mathematics was not so direct. He was forced to resort to the *regula falsi*, or rule of false position. The problem itself he set up by supposing that the ratio of speeds exceeded unity by just as much, *arithmetically*, as unity exceeded the ratio of distances. That is, the proportional excess of the inner planet's speed over the outer planet's equaled the proportional defect of the inner planet's distance. Thus

$$\frac{V_e}{V_m} - 1 = 1 - \frac{R_e}{R_m} \tag{4}$$

In fact, this equation, which is equivalent to the condition Kepler satisfied by trial with the *regula falsi*, reduces easily to equation (1). Kepler had made the same comparisons of ratios by subtraction here as in the analysis of periodic times and distances, and had managed to arrive at identical conclusions.

It is impossible to tell, so briefly is the theory sketched, just how the origin of the speed-distance theory was related to that of the equivalent period-distance theory. Kepler was apparently unaware that the two theories were identical until he saw that the results were the same. The period-distance theory had been based directly upon the perceived pattern in the Copernican parameters, while the speed-distance theory grew from his physical insight into this pattern. The more physical speed-distance theory, as stated and quoted above, seems as if it should correspond to an inverse proportionality of the usual kind, like (3); but Kepler interpreted his own words differently. As before, it is possible that comparison of various numerical procedures with

---

[11] *G. W.*, 1: 71: 36–39.

the desired Copernican parameters eliminated the alternatives to (4). I think this is less likely here than before, partly because of the computational difficulties of solving many similar problems iteratively with the *regula falsi*, and partly because that computation itself suggests a more likely explanation for the procedure in (4).

The rule of false position, or *regula falsi*, is a simple way of solving difficult problems iteratively. One assumes a value for the unknown quantity, then computes other quantities dependent upon it. If the problem has a unique solution, one will eventually obtain a contradiction, unless the assumed value was in fact correct. When the contradiction occurs, one adjusts the initial value and tries again. In well-behaved problems, comparison of the results from different trials permits one to converge on a correct solution rapidly, if tediously.

An initial assumption is needed. In his published example using the earth and Mars, Kepler took 694 as the assumed radius of the earth's orbit, where the orbit of Mars was normed to 1000. This value, 694, was in fact the one obtained from the period-distance theory earlier in the chapter. The *regula falsi* confirmed this value immediately, which should not surprise us, since we know that the value was calculated from a theory equivalent to the one being investigated here. Kepler himself simply remarked that, if the calculation had not confirmed the assumed value, one would have continued according to the *regula*. This remark leaves the impression that he had used the correct initial value merely to save a lot of computation in the book. This impression may be correct; narrative courtesy almost requires rapid convergence in a book.[12] Suppose, however, that one day he had been investigating some of the implications of the period-distance results, or of his physical speculations generally, and had used those results *verbi gratia*. The calculations of the speed-distance theory come naturally, just as in the example he published. The earth's orbit is (say) .694 that of Mars. Mars completes its revolution in 687 days, so at its normal speed it would traverse the earth's path in only .694 × (687 days), or 477 days. The earth makes it around in only $365\frac{1}{4}$ days, so it is faster than Mars in the ratio $477 : 365\frac{1}{4} = 1.306$. The proportional excess of its speed, .306, is precisely the proportional defect of its distance (1 − .694 = .306).

This coincidence, as it seems, is quite surprising when one encounters it in Kepler's example, as in the preceding paragraph. Why should he have looked further when the pattern seemed so obvious? There is really no coincidence at all, but to formalize the procedure in the above computation and show its equivalence to that leading to equation (1) was clearly beyond Kepler's powers in 1596. If he stumbled on the coincidence for one pair of planets he would naturally have tried it for other pairs; and of course, whatever initial values he chose, the *regula falsi* would have swiftly led him to the results of his period-distance theory. Even though his trial procedure had attracted his

---

[12] Kepler later worked out several iterations in Chapter 16 of the *Astronomia nova*, but there he was using the *regula falsi* twice in a double series of nested iterations.

attention by its ability to produce the first "coincidence," the other four would seem, on the face of them, to be independent confirmations of the proportions. Furthermore, the procedure made physical sense, in its rough verbal form.

No wonder, if he found it this way, that Kepler presented the speed theory in addition to the period theory. He was, however, sensible enough to recognize that the mutual corroboration of the two was too good to be real corroboration. They must be "identical in truth, and based on the same foundation,"[13] although he was not yet sufficiently skilled to understand why.

After devoting the next chapter to a not very helpful comparison of the regular-solid and motive-virtue theories, with one another and with the observed greatest, least, and mean distances of the planets, Kepler turned in Chapter 22 to consider the variation of the motive virtue in the course of a single planet's revolution. All of the planets moved on orbits eccentric to the sun; therefore, reasoned Kepler, the speed of each planet should vary in its orbit for the same reason the speeds of the different planets varied. This variation in the speed of a planet in its orbit, of course, is real. Kepler's abstraction of such an important fact from the Ptolemaic and Copernican theories, both of which combine regular motions so as to conceal the variation in speed, was of fundamental importance for his later career. Here he presented the result as a consequence of physics, rather than an implication of existing theory. Near aphelion the planet ought to move slower because of the weaker virtue at that distance, and similarly near perihelion it ought to move faster. The question then arose how one could conveniently measure the changing speed of a planet as it moved through its orbit.

Suppose that A in Figure 2 is the sun, and circle EFGH is the planet's orbit, centered on a point B eccentric to the sun. For comparison, draw also a circle NOPQ centered on the sun, and equal in size to the orbit. First imagining the planet to move with constant speed, one sees that near EF it is at greater than average distance from the sun, and therefore *appears* to move slower by an amount proportional to its excess distance there, compared to how it would appear on the concentric circle. The excess distance, near aphelion, is simply the eccentricity AB. But the planet does not move with constant speed: it is slower than usual around EF, again, according to the physical hypothesis, by an amount proportional to its excess distance over the mean. It thus appears to move as slowly as if it moved with *constant* speed in an orbit that deviated *twice* as far from the sun as does the real one. Such an orbit may easily be represented, by an imaginary circle IKLM, around a center D twice as far from the sun as the actual center of the orbit. Moreover, similar considerations at perihelion indicate that the planet, there both close and fast, would appear to move in this same imaginary circle IKLM.

And lo! Ptolemy, with scant explanation and to the amazement of his successors, had stated that the planets move in just this way, on an eccentric circle, but that they traveled as if in a circle of twice the eccentricity. Ptolemy—

---

[13] *G. W.*, 1: 72: 20–21.

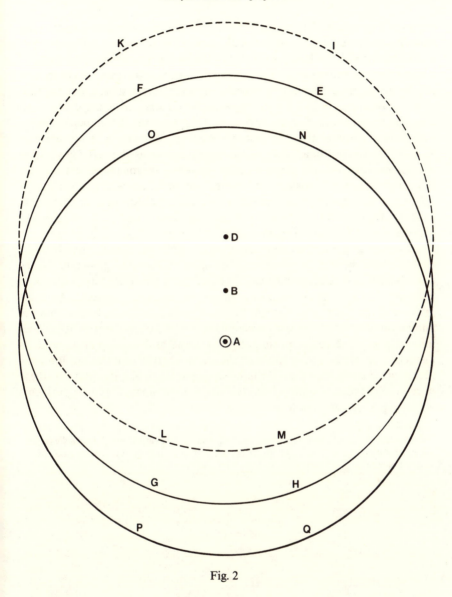

Fig. 2

and Copernicus, who rearranged the geometry of his model—had doubtless been unaware of the true physical reason why the planet moved in this way. That reason had lain hidden until, with an insight suggested by the innovation of Copernicus, Kepler had started to wonder about the distances and speeds of the planets.

Kepler's recognition of the real physical phenomenon of changing speed had triggered this insight, which guided his work through the rest of his life. At this crucial point (in 1596) his teacher Maestlin contributed invaluable technical support, explaining just what the model of Copernicus implied, and how that model supported Kepler's physical supposition.[14]

We have not much to say here of this, Kepler's first attempt to explain the motion of a planet in its orbit. The argument was really too casual to be treated as a demonstration. Essentially, however, it was the very same argument that would turn up again in Chapter 32 of the *Astronomia nova*, there advanced with geometry instead of hand-waving (below, pp. 63–66). In that book it introduced to astronomy Kepler's distance law, the foundation of his physical astronomy, known today chiefly in its mature guise as the "area law." Like the *Astronomia nova* demonstration, Kepler's early argument concentrated on the situation near the apsides, and played fast and loose with substitution of arithmetic for geometric progressions. He did not claim here (or there) any exact equivalence of the equant to his physical theory, and his result is in fact very accurate for orbits of small eccentricity, such as those of the planets.

Closing Chapter 22, Kepler confessed that there remained certain obstacles to the physics he had proposed. According to all accepted planetary theory (including that of Copernicus) the speed of the interior planets, Venus and Mercury, was not regulated by the model of Figure 2, and in fact did not increase as the planet approached the center. Kepler observed further that the anomalous motion of the earth remained to contradict his theories. Its annual motion had some cause, presumably the same as for the planets, and its distance from the sun varied, but all the experts agreed that this motion took place with constant velocity. If there was any truth in Kepler's physics, these were puzzles indeed. In the *Mysterium* he left them unanswered, as problems for an expert astronomer to solve.

---

[14] A. Grafton, "Michael Maestlin's Account of Copernican Planetary Theory," *Proceedings of the American Philosophical Society*, 117 (1973), pp. 523–550. The important letters are in *G. W.*, 13: 54–65, 108–119.

Chapter 3

Astronomia nova

In 1609 Kepler published the *Astronomia nova*, the record of a decade's intense labor. The full title of the work[1] proclaims that his new astronomy is causal, that it is a physics of the heavens based upon an examination of the motions of the planet Mars. This book is the pinnacle of pre-Newtonian astronomy, and points the way toward modern astronomy as established by Newton. Yet the material treated is scarcely that of a definitive treatise. No trace does one find, in this book, of the assurance with which Ptolemy had explained his predecessors' models, and his own, and no trace of the orderly and comprehensive arrangement of Ptolemy's *Almagest*. The *Astronomia nova* is instead the account of a trip into unknown territory. We enjoy today a comfortable familiarity with the end results, the area law and the ellipse, so that the journey narrated by Kepler seems even more circuitous than it really is.

To be sure, the exposition is marvelously indirect. Digressions, repetitions, and minute analyses of models already proclaimed incorrect make the modern reader despair of arriving at those two laws of planetary motion, if they are his goal. Kepler was not writing for the modern reader. He was writing for seventeenth-century astronomers, professional astronomers who knew quite a lot about planetary theory. Whatever they thought of Copernicus—whose work had profoundly unsettling implications outside their field of professional competence, in physics and theology—they could not have been prepared to accept Kepler's book, a book which replaced not only the old models of planetary theory, but the techniques of calculating from a model and the very concept of what the model was supposed to be. In recasting so much of a sophisticated and well-established science, Kepler knew he could expect resis-

---

[1] Astronomia nova *ΑΙΤΙΟΛΟΓΗΤΟΣ*, sev physica coelestis, tradita commentariis de motibus stellae Martis, ex observationibus G. V. Tychonis Brahe.

tance from its practitioners, his readers. He had to persuade them that his astronomy was not merely an acceptable way of computing planetary positions, but the true system of the world. This was why he led them on such a roundabout route to his theories of planetary motion. Certainly they had to reach that destination, but they also had to realize that there was nowhere else to stop. His strange diagrams and calculations and his peculiar theories were worth learning, if at all, only if the planets really moved in just the way he said. In view of the extreme novelty of its contents, the *Astronomia nova* was not so strangely written after all.

The book had a three-fold task. First, Kepler had to attack and discredit the old hypotheses, in their own terms. Armed with Tycho's observations, which were of an accuracy previously unattainable, he had to demonstrate that the models used by all astronomers were wrong, and that they could not be salvaged in any reasonable manner. Second, he had to develop his new astronomy, and show it to be accurate to the limits of the Tychonic observations. Finally, he had to argue that, compared with the alternatives, his theories were much more likely to represent what was actually taking place in the heavens. For the first and last of these reasons, Kepler reintroduced physical argument to astronomy, and thereby shifted the overall emphasis of his book from the mathematical representation of observations to the determination of how and why the planets, huge, physical bodies, moved through the heavens.

We must suit our discussion of physics in the *Astronomia nova* to its use in the argumentative plan of the book. Through the first thirty-one chapters, nearly half the book, Kepler's physical speculations are just that: speculations, incidental and rather vague. This is not to deny their importance, for it was often precisely the point that certain models began to look improbable upon merely raising the physical questions, in however general a fashion. Still, Kepler intended by these early physical arguments to plant suspicions, not to propound a theory. Rather than attempt to fasten on the precise meaning of these abstractions, we shall simply give an account of them and see how Kepler *used* them in his critique of models previously employed by astronomers. The analysis in these chapters was subtle and, except for the physical arguments, highly technical, but we shall attempt to describe it in outline.

In these early chapters, Kepler made fundamental improvements in the theory of the first and second inequalities of longitude, and in latitude theory, and showed that Tycho's observations would not fit any theory along the traditional lines. He therefore set out to find the correct theory. In the course of his search he engaged in the extended physical discussions of Chapters 32–39, beginning to fashion a physical theory of planetary motion. Through the remainder of the book, he took up this theory from time to time, reconsidering and revising it. The later chapters, where the physics is connected with the remarkable discoveries formulated in the first and second laws, will bear more intensive analysis than the early discussions of physics. Alone, Kepler's physics is something of a curiosity; combined with his astronomy, it is extraordinarily interesting.

Preliminaries

The principal business of Part I of the *Astronomia nova,* the first six chapters, was the replacement of the mean sun by the true sun in planetary theory. In the meantime, well-known preliminaries had to be stated. The observed motion of a planet is marked by two periodic irregularities, called "inequalities" or "anomalies." For Mars, the "first" anomaly arises from the motion of the planet itself. We, along with Copernicus and Kepler, attribute the "second" anomaly to the observer's location on a moving earth. Tycho had ascribed it to the sun's motion around the earth, carrying with it the entire Martian orbit; Ptolemy, to the motion of the planet on an epicycle. The three second-anomaly models are identical mathematically. Kepler demonstrated this equivalence, which freed him from the need to take constant notice of the Ptolemaic and Tychonic models in the remainder of the book. (He did, however, sketch out the Ptolemaic and Tychonic forms of his analysis in Parts I and II, both to demonstrate the independence of his mathematical results from any particular world-hypothesis and to honor a deathbed request of Brahe.[2])

The first inequality could also be treated by equivalent models. The simplest supposition was that Mars moved on a circle eccentric to the sun. Alternatively, one could use a concentric circle, with an epicycle rotating in the opposite direction from the concentric. A much better approximation, discovered by Ptolemy, was to employ an equant circle. The equant model supposed the planet to move on an eccentric circle, but to regulate its speed so as to maintain constant angular velocity about a point other than the center of this eccentric, namely the center of an imagined "equant circle." Copernicus, perhaps following the fourteenth-century astronomers of the Marāgha school, had achieved virtually the same effect as the equant model by using an extra epicycle.[3] For reasons that will soon become clear, Kepler preferred the equant to Copernicus's epicyclic model for the first anomaly, to represent the proper motion of Mars. He showed in Chapter 4 that the two models were so closely equivalent that astronomers who preferred the Copernican epicycle could accept the results of Kepler's analysis of equant motion.

Aside from these formalities, Kepler's point in the first six chapters was that the sun itself, the physical body of the sun, must play the major role not only in general cosmology, but in the detailed geometrical constructions of planetary theory. To make this point he needed to press upon his readers the physical implications of planetary theory, and thus to make them think about what was really happening in the skies. For relatively simple reasons, these implications had changed radically in the closing decades of the sixteenth century: there were no longer any solid spheres.

Ptolemy's *Almagest* had described only the geometry of planetary models.

---

[2] *G. W.,* 3: 89: 7–11.
[3] M. Maestlin, in *G. W.,* 1: 138; E. S. Kennedy and V. Roberts, "The Planetary Theory of Ibn al-Shatir," *Isis* 50 (1959): 227–235; O. Neugebauer, "On the Planetary Theory of Copernicus," *Vistas in Astronomy* 10 (1968): 89–103.

The physical structures underlying these models were agreed upon, however. Rigid, transparent spheres, properly nested, could reproduce in three dimensions all of the elaborate motions depicted in the geometrical diagrams. In Chapter 2, Kepler presented these solid-sphere models, first in a simple form described by Aristotle, and then according to the more sophisticated Ptolemaic versions which had been explained by Peurbach in his *Theoricae novae planetarum* (first published 1472). Aristotle's model had employed only spheres concentric to the earth, and was much too crude for astronomical use. It did exhibit both the strong and the weak points of these early physical models, in features which were carried over into Peurbach's more elaborate models.

We must restrict our discussion to general characteristics of the solid-sphere theories. A solid-sphere model, because of the close nesting of the spheres, needed only to be moved. As the spheres rotated in place, carrying the other spheres that were interior to them and carried by those exterior, their rigidity constrained them to the desired geometrical paths. There remained, of course, a need for something, a 'motor' or mover, to actually produce the motion. One could not by studying the motion learn very much about these movers, for their action was exceedingly simple. It seemed safe to conclude that the planetary movers were not material, since they acted through all eternity. Aristotle, as Kepler reported, had attributed to them understanding and will, to direct the motion; and had attached to each mover a motive soul to assist in imparting motion to the resisting matter of the spheres. Kepler evidently thought this a plausible enough arrangement, on the face of it. He observed that the motion of the spheres in such a model appeared to involve, first, an actual force, producing the motion and determining the rotational period by the ratio of its constant strength to the resistance of the matter; and second, an ability to guide the force in the correct direction. The former task he thought appropriate to some kind of animal faculty, the latter to a property of understanding or remembering. (One should bear in mind these attributes. When Kepler attributed a soul, *anima*, to a heavenly body he was trying to account for some manifestation of constant but mindless motion. When he considered the possibility of planetary minds, it was because motion appeared to be directed in a way that required either calculation or memory.) Solid-sphere models, in sum, were constrained by their structure to follow the correct paths, but had to be moved by something whose nature remained more or less a mystery.

Hypotheses including epicycles or eccentric circular paths required more complicated arrangements of the spheres; but Peurbach and others had devised suitable combinations. As a Copernican, Kepler discussed these more elaborate structures as representations of the first anomaly, although Ptolemaic astronomers had more often used the epicycle configuration for the second anomaly. In a simple epicyclic model (Figure 3a), one visualized the epicycle as a small rotating sphere, carrying the planet somewhere near its surface; with the whole enclosed in, and carried by, a spherical shell that was rotating about its own center. In the equivalent simple eccentric model (Figure 3b), the eccentric was a shell at least as thick as the planetary body, and

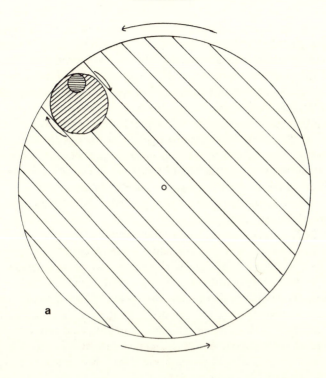

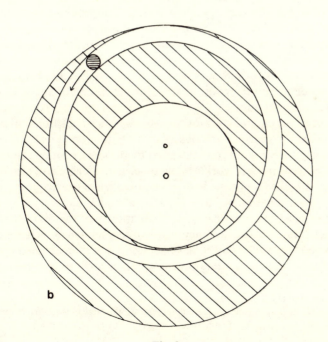

Fig. 3

carrying it. This shell was eccentric because it moved in the off-center cavity of a larger shell. In turn, the eccentric shell contained another, smaller shell, whose interior cavity was displaced back so as to be centered on the center of the world. This nest of three shells permitted the middle one, carrying the planet, to rotate around one center, while the exterior surface and interior cavity of the nest were aligned with the center of the world.

The simple epicyclic and eccentric models both moved the planet according to the same admittedly oversimplified geometric model of the first anomaly. Neither required anything in the way of directing minds, motive souls, etc., beyond what was in the Aristotelian model, for the nested spheres guided all the motions along their correct paths. As Kepler put it, the movers of the spheres "are directed by material necessity, that is by the disposition and contiguity of the spheres."[4]

The full Ptolemaic first-anomaly model required the planet[5] to maintain constant angular velocity about the center of the imaginary equant circle. This implied, of course, that the actual shell carrying the planet varied its velocity, speeding up and slowing down so that the angular velocity about the equant center remained constant. In this equant model, therefore, the solid-sphere representation lost much of its elegance, for the necessary variation in speed was left unconstrained by the physical construction of the model. Fortunately, the failure of solid-sphere models to encompass equant motion had been resolved by the Marāgha astronomers and by Copernicus with an epicyclic model that mimicked equant motion.

The solid-sphere planetary models were really not so much a celestial physics as a reason why no one had ever developed any celestial physics. Greek astronomers had accounted for the motions of the planets geometrically, and as long as it was possible, or appeared possible, simply to transform the geometry of the hypothesized motions into an equivalent geometry of solid shapes, which real matter could assume, further speculation remained outside the domain of astronomy.

Kepler, however, could no longer rely on "the disposition and contiguity of the spheres." Tycho Brahe's observations of the comet of 1577 had shown convincingly (to Kepler, at any rate) that there were no solid spheres to guide the motive souls in their work.[6] Some other guide must exist, and if Kepler's search for it carried him outside the traditional realm of astronomy, he simply had to expand that realm. The solid-sphere models had long coexisted with mathematical astronomy. They were compatible with the geometrical models for the very direct reason that they too were geometrical models. Once Kepler had realized that physics must be something other than geometry, he ran up against the problem of finding principles which were plausible in their own

---

[4] *G. W.*, 3: 68: 40–41.

[5] Actually, the center of the epicycle representing the second anomaly. We shall frequently follow Kepler in discussing Ptolemy's first-anomaly model as if it were heliocentric.

[6] *G. W.*, 3: 69: 1–3.

right, but mathematically equivalent to the planetary models of astronomy proper.

In this task he was almost entirely on his own. Contemporary physics was not going to offer any help, and he was essentially left free to speculate about the kinds of things which were required to impose order on the motion of planets traveling "in pure aether, just as birds in the air." [7] To start with, a power capable of moving the planet was necessary, as it had been in the models of Aristotle and Peurbach. Although he never wrote systematically about motion in general, Kepler always assumed, and more than once stated, the principle that bodies would remain at rest unless moved by some force. Throughout Kepler's physics we shall encounter consequences of this fact, that the principles he assumed did not include the modern or "Galilean" law of inertia.

Ignorant of inertia, Kepler had no particular insight into the "causes of motion." It was the other task, that of directing the motive force, which he particularly sought to understand, and it was this other task whose aspect had decisively changed with the dissolution of the spheres. Consider the simple model of the first anomaly, as represented either by an eccentric or by an epicycle. The planetary body had to follow a circular path—either the eccentric or the epicycle—of which neither the center nor the perimeter was distinguished in any way from the surrounding "aetherial aura," [8] the stuff filling the translunary world. More specifically, in the eccentric model the planetary mover had to calculate its distance from the sun, perhaps by examination of the size of the solar disk, and from that distance to determine the direction in which it should travel to remain on the eccentric circle. This required, among other things, a knowledge of the planet's eccentricity and a certain amount of mathematics, and was a rather more complicated task than Kepler wished to assign his planetary mover. The alternative, which was that the mover contrived to maintain constant distance from the eccentric center, he did not think possible. The center was an utterly vacant point, from which it would be impossible to measure one's distance.

The simple epicyclic model was even worse. Here one had to imagine a disembodied power or "motive virtue" [9] traveling around the central body in a circle; and a second virtue in the body of the planet which was able to

---

[7] G. W., 3: 69: 4.

[8] *Aura aetheria*: Kepler's favorite phrase for whatever it was that filled the heavens. The word *aura* can refer to the upper air or heavens; Kepler's use presumably derived from some such sense of the word. I do not know what properties he thought it to have, aside from transparency and thinness.

[9] *Virtus motrix*. When I am discussing a specific passage or argument, as here the one at the end of Chapter 2, I shall try to be careful with terms, rendering *virtus* as virtue, *vis* as force, *anima* as soul. Kepler often used *virtus* and *vis* indiscriminately, although more properly a virtue is a capacity for exerting a force. In his mature work, Kepler attributed souls to heavenly bodies only when they appeared to move of themselves, that is without being moved by anything external. Such self-motion always consisted of rotation in place.

perceive the first—although there was nothing material to perceive—and to carry the planetary body around it, again in a circle. Clearly any theory in which the motion of the planet was guided from the planet itself faced some very tricky problems.

Kepler thus used some admittedly vague speculations, concerning the difficulty of controlling a planet's motion with information available at the planet itself, to suggest that at least part of this task took place elsewhere: presumably, therefore, at the central body. In the Copernican and Tychonic systems, the central body was the sun, which however had no function in planetary theory beyond sitting near the middle of things, illuminating. Planetary models were instead constructed around the "mean sun," the place the sun would appear if its apparent motion were uniform, which in Copernican theory coincided with the center of the earth's eccentric circular path. Now, if the sun participated in directing the planetary motions, must it not appear in any hypothesis of those motions? This was the conclusion Kepler wished to draw, the conclusion he had anticipated in Chapter 15 of the *Mysterium* (p. 11, above). Being a very careful astronomer, and wanting particularly to appear such in a book filled with novelties, he first considered its consequences in some detail. The discussion afforded him the opportunity of bringing forth some important evidence for his theory that a force moving the planets was seated in the sun.

As mentioned above, the simple eccentric or epicyclic model Kepler discussed in Chapters 2–3 could not accurately represent the first anomaly. Ptolemy had employed an equant, whereby the center around which angular motion was uniform differed from the center of the circular path, so that the actual movement around the circle was faster at perigee than at apogee. Copernicus, as Kepler noted in Chapter 4, had rejected the equant hypothesis because of its physical absurdity as a model for motion of rigid heavenly spheres. Indeed, to move a rigid sphere according to the equant model would either distort the supposedly-perfect sphere, or require the speed of rotation to vary. Varying the speed introduced the same kind of difficulty Kepler had discussed regarding eccentric circular motion through the aether: the planetary mover would either need to pay careful attention to the equant center, which was invisible, or it would have to make fairly complicated calculations based on abstract knowledge. Copernicus, as we have said, had replaced the equant with a nearly-equivalent epicyclic mechanism that could be achieved with uniformly rotating spheres.

Kepler would have concurred in this epicyclic model, he confessed, had the motive virtues been in charge of solid spheres instead of bare planets. In the actual case, the complex Copernican mechanism was difficult to conceive, while he found a certain physical elegance in equant motion. To him the equant model now seemed "nothing other than a geometrical shortcut for computing the equations from a completely physical hypothesis," [10] although

---

[10] *G. W.*, 3: 74: 10–11.

he acknowledged that Ptolemy had been entirely unaware of its physical basis. More precisely, Kepler preferred the equant as a convenient and transparent way of representing what was for him the critical phenomenon: that the planet moved swiftly when near the sun and slowly when distant from it. This was a consequence of both first-anomaly models, but more evident in the Ptolemaic. The Copernican model, besides concealing the variation in speed behind a combination of uniform motions, would have required an intolerable amount of "mental" activity to control the motion. The Ptolemaic equant, on the other hand, by openly displaying this variation, encouraged Kepler's attempt to locate an impelling and guiding force in the sun. If only some way could be found to explain the planet's approach to and withdrawal from the sun, the variation of speed would be easily understandable as a consequence of the weakening of some solar force with distance from its source. Variation in the planet's speed, which in earlier astronomy had been a blemish to be ignored or concealed, singled out the sun now as the heavenly body which had to be somehow involved in moving the planet.

The importance for Kepler (and for astronomy) of this physical interpretation of Ptolemy's equant cannot be overemphasized. It does appear ironic that Kepler so quickly deserted the new first-anomaly model, of which Copernicus had been so proud (and which Tycho himself, destroyer of the spheres, had preferred), and had returned to the older equant. His "return," however, was to an equant of quite different significance. Ptolemy, no doubt after elaborate and unreported analysis, had separated the center of the circle from the center of equal motion, in order to adjust the distances of his epicycle from the earth. (He had to alter these distances, according to the *Almagest*, to account for the greatest elongations of Venus when the epicyclic center was a quadrant from its apogee [X, 3], and to account for the times and arcs of retrogradation of the superior planets [X, 6].) His equant model was a geometrical expedient, uncannily accurate in the light of modern analysis, but in context possessed of no further significance than its effect of altering certain distances in the simple model, while leaving the equations of center almost unchanged. The model did imply that the planet sped up and slowed down, but this implication had been unsought and unwanted. The angles in the equant model gave the right equations for the first anomaly, and the distances gave the right equations for the second anomaly, and these two things were what mathematical astronomy required. Changes in the actual speed of motion were irrelevant, if awkward, in Ptolemy's astronomy, for one did not normally need to deal with the actual speed.

After the work of Copernicus and Tycho, Kepler saw all of these things differently. It was certainly still important that the model gave the right equations. In the Copernican system, however, the question was no longer one of an epicyclic center circling the earth, but of a planetary body circling the solar body. Since Tycho's work, it further seemed that the one body circled the other through empty space. Whatever moved the planet moved it continually faster as it approached the sun, and continually slower as it withdrew.

It was the *monotonic* change of speed with distance which was obvious in the equant model, but not in the alternate epicyclic model. This was exactly the sort of relationship Kepler needed, to analyze the physics of a planet moving through space. The equant was a ready-made foundation, designed for accurate equations but marvelously amenable to physical interpretation. Upon its monotonic variation of speed with distance from the sun he built his physical astronomy.

Before turning to the mathematical preliminaries of Chapters 5 and 6 we should remark upon Kepler's use of "minds" in his discussion of physical problems. This question is more complex, and for our purposes more important, than the parallel question regarding souls. In these early chapters of the *Astronomia nova*, souls furnished the motion and minds the guidance. It is no simple matter to say how far Kepler believed in the actual existence of minds in the heavens, or of what kind he thought them to be. Physically, Chapter 2 was hypothetical, dealing with the minds guiding solid-sphere models and with the eccentric circles and epicycles which remained when these spheres were denied, so from it we cannot draw any firm evidence. What is clear is that when Kepler spoke of a mind in the heavens he was considering the problem of *controlling* motion; more specifically, the problem of obtaining sufficient information to constrain motion into the regular path that was observed. However one supposed a planet to be moved, and whatever one thought to be the cause of its motion, the regularity of that motion must have originated somewhere and must by some path have been transmitted to the planet. Not even an intelligent mind could direct motion along a circle unless it perceived some pattern that told it where the circle was. Kepler was willing to consider the possibility of minds unknown to us, but did not consider that they might perceive and act upon patterns hidden to the human mind. Physical astronomy would have been quite unattainable if planetary motion depended upon information inaccessible to the astronomer.

Kepler did believe it possible to understand the arrangement and workings of the universe. In part such understanding had to come from the rational and aesthetic patterns he had studied in his *Mysterium*, and would study again in the *Harmonice Mundi*. The incomprehension and lack of sympathy of later generations have left his work on these cosmic harmonies in the obscurity he foresaw.[11] Nor can we expand our present topic to include much of it, although archetypal relations enter into the physical lunar theory of the *Epitome* and we will examine them there (pp. 190–195, below).

Alongside the rational organization of the universe were the physical bodies themselves, the sun, planets, and stars. To the extent that perceivable relationships between these bodies exhibited the same regularities as their motion, Kepler presumed that the regularities of motion somehow arose from the regularities of configuration. Planets moved slower when farther from the sun, so whatever moved them must evidently relate to their distance from the sun.

---

[11] Proemium to Book Five of the *Harmonice Mundi*, in *G. W.*, 6: 289–290.

Where, on the other hand, a regular pattern of motion was accompanied by no corresponding regularity from which it could arise, he suspected that the motion itself was not understood. His critique of epicyclic motion was of this kind: there simply was no constant relation by which the planet's constant distance from the epicyclic center could be known, and hence he doubted that there could be any means of maintaining the constancy.

The best part of Kepler's physics, throughout his life, was the persistance with which he analyzed the information implicit in a regular pattern of motion, and sought to trace this regularity back along some path of causality to its source. He never had much success explaining what moved the planets, for as we know now this was a false question. On the question of *how* the planets moved, he began with the assumption that they were acted upon by a force which must originate from a physical body. In the end he accurately described how they moved, correctly analyzed this motion in physically-meaningful coordinates, and gave a plausible, though completely wrong, explanation of how the motion arose.

The last two chapters of Part I of the *Astronomia nova* were devoted to an examination, in detail, of the changes which ensued in all three world-systems if one replaced the mean sun (or its equivalent) by the true sun. The problem was this.[12] In order to construct a model for the first anomaly, the inequality proper to the planet, one needed to know where the planet was with respect to the sun or the earth. (Kepler spoke in general terms of "the center from which the eccentricity is measured.") One ascertained planetary positions as they would appear from the sun by the simple expedient of waiting until the earth lay directly between the calculated position of the sun and the observed position of the planet. Normally one had to interpolate between several observations taken at about the right time. Real or interpolated observations made in this way were called acronychal.

The procedure was simple enough, given a rudimentary solar theory, and permitted astronomers to "observe" the planet from whichever center they deemed appropriate. Copernicus had taken this center to be the mean sun, which for him was the center of the earth's path. Kepler believed for physical reasons that one should measure the eccentricity from the sun itself. The choice makes a real difference, as we can see in Figure 4. Let C be the center of the eccentric circle of Mars, A the sun, and B the mean sun. Copernicus thought that F, the point farthest from the mean sun, was the aphelion; Kepler chose D, the point farthest from the true sun. Now if Mars moved with constant velocity on its circle—it does not—the disagreement would be merely verbal. Either way one could agree that the planet was at, say, P; and it would be of little moment whether one said it was distant from (Copernicus's) aphelion F by arc FP, or distant from (Kepler's) aphelion D by arc DP.

---

[12] We shall restrict ourselves to heliocentric models with an equant, although Kepler sketched out most of these arguments in geocentric and Tychonic form also, and in the Copernican epicyclic first-anomaly model.

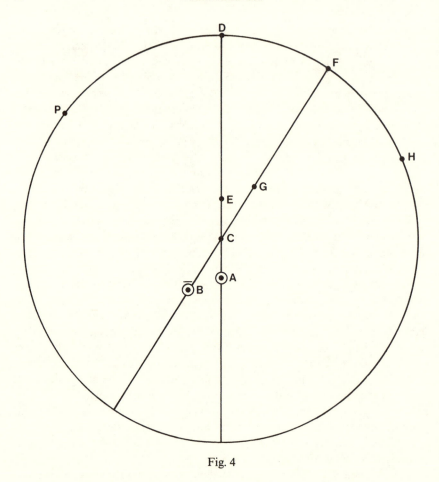

Fig. 4

Mars did not move with constant velocity on its circle. It moved slowest at aphelion: at F if, like Copernicus, one thought aphelion to be the point farthest from the mean sun, but at D if aphelion were taken from the true sun. This variation in velocity had important consequences. To begin with, if the apsidal line really passed through the sun, no astronomer who placed the apsidal line through the mean sun could locate the planet's path correctly. He would find the observed motion to be slower to one side of his apsidal line than the other. Specifically, the planet would take longer to move from F to D than it had taken to move an equal arc from H to F.

In order to see the predicament of an astronomer who had thus misplaced his apsidal line, let us examine the effects of combining a real change in the planet's velocity with a further, optical change in the apparent velocity due to changes in the planet's distance from the observer. (Whatever we speak of a planet's speed or velocity without qualification, we mean the physical or linear

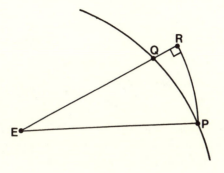

a

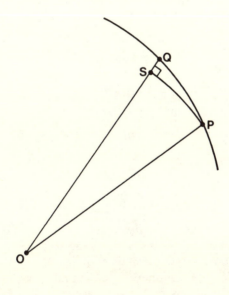

b

Fig. 5

speed, distance traveled per unit time. When we mean angular or apparent velocity, we shall carefully distinguish it as such.)

In Figure 5a, PQ is a section of the planet's path, and PR a section of an equant circle, that is, a circle centered on the equant point E. The path element traversed in a short time Δt is given by

$$PQ = PR/\cos(RPQ)$$

$$PQ = PR/\sin(EPQ)$$

But PR is a circular arc given by

$$PR = EP \cdot \Delta t$$

Hence

$$PQ = EP \cdot \Delta t / \sin(EPQ)$$

$$\frac{PQ}{\Delta t} = \frac{EP}{\sin(EPQ)} \tag{5}$$

Since Δt represents the time taken to travel from P to Q, $PQ/\Delta t$ is the planet's (linear) velocity. We see, therefore, that the planet's velocity is proportional to its distance from the equant, divided by the sine of the angle between the radius from the equant and the tangent to the path.

Consider now the angular or apparent velocity as seen from some other point O (Figure 5b). PQ is the same arc of the planet's path, while PS is an arc of a circle centered on the observer at O. The observed arc PS is then

$$PS = PQ \cdot \cos(SPQ)$$

$$PS = PQ \cdot \sin(OPQ)$$

The *angle* subtended by this arc, let us call it ΔP, is then

$$\Delta P = PS/OP$$

$$\Delta P = (PQ \cdot \sin(OPQ))/OP$$

So that the angular velocity is

$$\frac{\Delta P}{\Delta t} = \frac{PQ}{\Delta t} \cdot \frac{\sin(OPQ)}{OP}. \tag{6}$$

The angular or apparent velocity equals the linear velocity divided by the distance to the observer, times the sine of the angle between the radius to the observer and the tangent to the path.

We may unify all this by speaking of R_E and R_O as the radii to the planet from the equant and observer, respectively, and Θ_E, Θ_O as the respective angles these radii make with the tangent to the planet's path. (These angles are the complements respectively of the physical and optical parts of the equation of center.) We then have

Velocity proportional to

$$\frac{R_E}{\sin(\Theta_E)} \tag{5'}$$

Apparent velocity proportional to

$$\frac{Velocity \cdot \sin(\Theta_O)}{R_O} \tag{6'}$$

Apparent velocity proportional to

$$\frac{R_E \cdot \sin(\Theta_O)}{R_O \cdot \sin(\Theta_E)} \tag{7}$$

Roughly speaking,[13] the apparent velocity is proportional to distance from equant divided by distance from observer. It is nearly true, then, that the apparent velocity will be least where R_E is smallest and R_O greatest. In a normal equant model, E and O lie on a diameter, the apsidal line, and this criterion is precisely satisfied at aphelion. Note, however, that equations (5')–(7) and the small-angle approximation to (7) do not assume E and O to lie on the same diameter.

In Chapter 5, Kepler analyzed the more general equant model, where the equant and observer do not lie on the same diameter. No one had ever proposed such a model. Copernicus, however, had based his model upon motion as observed from the mean sun (B in Figure 4); while according to Kepler's physical analysis that motion had really been uniform about an equant E which was not on the same diameter as B. More to the point, Tycho's assistant Longomontanus had also placed his apsidal line through the mean sun, and had developed a model which agreed with all of Tycho's acronychal observations of Mars, to an accuracy of 2'. If Kepler's physical astronomy, which demanded that the sun itself be used in place of the mean sun, was to be convincing, he had to show how Longomontanus had attained such accuracy from erroneous principles. He did not want to go so far as to give the impression that the two placements of the apsidal line were observationally indistinguishable; he only wanted to show that the use of the mean sun could give good results in a conventional theory.

To treat these problems, Kepler had to determine how closely an equant model with the wrong apsidal line could mimic the apparent motion from the correct equant model. As we have seen, no astronomer could use the right orbit[14] with the wrong apsidal line, for the physical inequality of the planet's motion would be asymmetrical to that apsidal line, and this asymmetry would have directly contradicted the model. What, then, would the astronomer think he was observing? Equation (7), above, shows that the observed motion would be slowest at a point, if one existed, which was closest to the equant and at the same time farthest from the observer; but there is no such point. If we connect the misguided observer at B with the equant at E (Figure 6) and continue the line to the circle at K, then D is the point closest to the equant, F the point farthest from the observer, and K a point which seems to roughly minimize the ratio of equant distance to observer distance. An astronomer who thought the apsidal line passed through the mean sun B would observe the slowest apparent motion (from B) in the direction of E and K, and hence

---

[13] That is, for orbits of small eccentricity, where Θ_E and Θ_O are always close to 90°, and their sines always very close to 1.

[14] The word "orbit" is somewhat anachronistic in discussion of pre-Keplerian astronomy. Kepler, however, was quite clearly studying an orbit: not a solid-sphere model, not a mathematical hypothesis, but a trajectory of a planet moving through space, acted upon by physical forces.

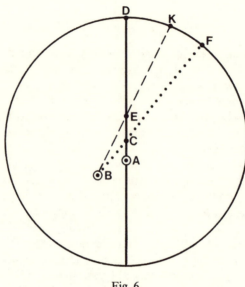

Fig. 6

would unwittingly direct his apsidal line through E, the true equant center.[15] His eccentric circle, however, would be different from that shown, for he would assume it to be centered on a point between the mean sun and the equant, on segment BE. His model, then, would be that dotted in Figure 7.[16]

Figure 7 represents the error which ensued from the practice of modeling the first anomaly of Mars around an apsidal line passing through the mean sun. Mars traveled the solid orbit, while astronomers (other than Kepler) thought it to travel the dotted orbit. The genesis of their error was innocent enough: it was the assumption made by Ptolemy, following the original idea of Apollonius, that the planet moved uniformly on the epicycle responsible for the second anomaly. This assumption was perfectly reasonable in geocentric astronomy.[17] When the Copernican transformation replaced all the Ptolemaic epicycles by the single eccentric circle of the earth, a great many new questions arose, and one of them was why the other planets varied their

[15] Kepler's argument does not require that this be the best placement; merely that it give results good enough to account for the model of Longomontanus. Incidentally, accurate computation, using (7) and the Tychonic parameters employed by Kepler, shows the point of minimum apparent velocity to lie slightly less than 43′ east of point E, as seen from point B.

[16] We have not yet shown that the equant point for the model based on the mean sun B would *coincide* with the former equant E, only that it would lie in the direction BE. Coincidence of the equants follows from the fact that both models agree upon two points of the orbit, marked U and V in Figure 7. The time taken to travel between these points must be agreed upon, so the angle UEV, and hence the location of E upon the dotted apsidal line, is fixed.

[17] Even Delambre, who usually was excessively critical of Ptolemy, absolved him of any fault on this point (1: 399).

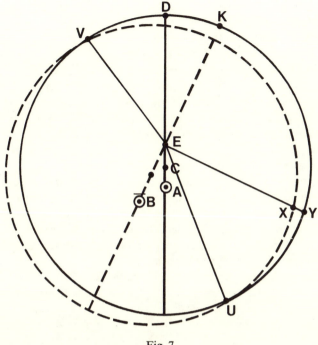

Fig. 7

velocity symmetrically with respect to the center of the earth's eccentric. Like most of the astronomical questions raised by the Copernican revolution, this one could neither be properly formulated nor answered until the solid spheres were cleared out of the way and the whole system reconceived as material bodies interacting through physical forces. The physical reinterpretation of Copernicus's theory, and the realization of its heuristic power, were Kepler's first solid accomplishments, and, one suspects, the easiest for him to attain. He was never blinded by traditional views on cosmology, having apparently discarded most of them before he attained competence as an astronomer. Yet we must not underestimate the importance to astronomy of these achievements, nor the originality required to detect and eradicate, among the complications of planetary theory, the parts that had become arbitrary and irrational in the newly discovered solar system.

At this point, however, Kepler's argument was simply that the error which astronomers made, in observing the first inequality as if from the mean sun, and in running their apsidal lines through the mean sun, could scarcely be detected—so long as they compared their model only to observations made as if from the mean sun. At any given time the two points on the two models of Figure 7 which lie on a line from the equant E are quite close, so long as they are seen end-on, from points near the center, such as the mean sun B. Kepler found the greatest angular discrepancy, which occurs around points

X and Y in Figure 7, to be less than $4\frac{1}{2}'$, using the Tychonic parameters for the (dotted) Martian orbit.

Was the error observationally undetectable, then? Kepler later treated Tycho's observations as if they were accurate within a couple of minutes. Two points should be made here. First, the observable error amounted to 4' or more only in one particular direction toward the beginning of Gemini. Acronychal observations of Mars in other parts of the zodiac would be liable to much smaller errors.[18] Second, as Kepler pointed out,[19] an astronomer by a slight adjustment of parameters could likely reduce the error still further. Anyone working with a particular group of observations tends to derive parameters especially suitable for those observations.

But these were relatively minor questions. The real difference between a planetary model built on the mean sun and one built on the sun itself was hidden until one tested the model outside of opposition to the sun: that is, with observations qualitatively different from those used to construct it.

The reason for this curious fact is easy to see. As we have already remarked, the two models in Figure 7 predicted, at any given time, two positions on the same line from the equant center E. So long as one observed these positions as if from the sun, mean or true, the difference between them was little: X and Y have nearly the same longitude as seen either from B or from A. They differ greatly, however, in distance from the sun. Differences in distance must be seen from the side to be perceptible. One built a first-anomaly model with acronychal observations (where the earth lay directly between the sun and the planet) because these gave heliocentric longitudes unconfounded by distance. This very fact meant that the model contained no observational information about the planet's distance from the sun. The model imposed distances *a priori*, by requiring the planet to trace a circle centered between the sun and the equant.

In Chapter 6 Kepler therefore considered the two circles of Figure 7 as seen from various points on the earth's orbit, rather than from the sun or mean sun. From this new perspective the shift in the location of the circle became visible. At the most opportune moment for seeing it, the difference amounted to an angle of over a degree, fourteen times the error at opposition and easily observable by any standards. Failure to place the planet's apsidal line through the true sun could have had disastrous results in Copernicus's astronomy. The only reason these had not been detected was that Copernicus, following the *Almagest*, had modeled the proper motion of the planet from acronychal observations, and these had not been sufficient to detect the erroneous distances in his theory. It was necessary to look at the distances from the side. This lesson Kepler remembered.

The argument thus far, we should emphasize, had been hypothetical. No

---

[18] Tycho actually did have an acronychal observation, from 1580, of Mars in the beginning of Gemini. I have not attempted to investigate the discrepancies between Kepler's theory and that of Longomontanus.

[19] *G. W.*, 3: 84: 12–16.

observations had been produced. Kepler had shown mathematically that one could construct a model for the first anomaly of Mars, with the apsidal line passing through the mean sun, that would represent acronychal observations quite well, even if the real apsidal line passed through the physical body of the sun instead. This explained how the model of Longomontanus performed so well. He had further shown that such a model could be found out as erroneous by looking at the *distances* of the planet from the center of its orbit. He had argued, on the basis of physics rather than astronomical observations, that in the absence of solid spheres to carry the planet, the regular shape of its motion, and particularly the systematic variation in its velocity, could only be explained on the assumption that the sun played an important role in guiding its motion.

Imitation of the Ancients

Kepler gave to the second part of his book the title "Concerning the first inequality of the planet Mars, in imitation of the ancients." It dealt with the first anomaly of Mars, the anomaly due to the planet's own motion; and it imitated the ancients in using models built from the uniformly rotating circles which Ptolemy had employed. Kepler took the ancient theories and generalized them, producing (for the first time since Ptolemy) a major improvement in the accuracy of the first-anomaly model, while still adhering to the principles of the ancient science.

In order to make this advance Kepler needed Tycho's observations. These far surpassed any that had ever before been accumulated, in accuracy and— equally important—in quantity. He began with the tedious work of reducing some of these observations to the form he required. This was necessary on two accounts. He had found, first, that Tycho's assistants had computed erroneously the ecliptic longitudes corresponding to the observed planetary positions. Second, they had reduced the observations to times of mean opposition, so that they gave the longitude of Mars as seen from the mean sun. Kepler needed, according to his "preconceived opinions,"[20] longitudes as they would be seen from the sun itself. That is, he needed to calculate the planet's position at the time of true opposition to the sun. His preconceived physical opinions required that the theory be built around the physical body of the sun.

Let us look more closely at the first of these problems. The orbits of the planets are inclined to one another at small angles. In particular, Mars does not travel in the ecliptic, the plane of the earth's orbit. Both Ptolemy and Copernicus, however, had treated the planet's motion in its orbit as motion purely in longitude, a coordinate measured in the ecliptic, and then had computed the planet's latitude separately, from separate theories. This pro-

---

[20] *Secundum praeconceptas et in Mysterio meo Cosmographico expressas opiniones. G. W.,* 3: 109: 40–41.

cedure involved an approximation, in that the planet's projection onto the ecliptic was assumed to move exactly as the planet moved in the plane of its orbit. In actuality, the projection moved a little slower than the planet around the nodes, and a little faster around the limits. To be strictly correct, then, the first anomaly should be modeled on the actual (inclined) path, for this was where the planet itself moved.

The difference was a very small one, and in planetary theory it had rightly been neglected by astronomers prior to Tycho. That astronomer, anxious to retain every benefit from the accuracy of his observations, had insisted on applying the small correction in his planetary theory. Unfortunately, his assistants[21] had taken this refinement too far. When calculating the moments of opposition, to obtain acronychal positions of Mars, they had found the time when Mars was as far from one of its nodes as the sun was from the opposite node. Kepler pointed out in Chapter 9 that this was a mistake. One wanted to know the heliocentric longitude of Mars, and longitude was an ecliptic coordinate. One therefore needed the moment when Mars's projection onto the ecliptic was exactly opposite the sun: at this moment the second anomaly of longitude (which was due to the motion of the earth, or for Tycho the sun, in the ecliptic) had vanished. The ecliptic position of the planet at this moment could then be projected back into its proper plane, and the angular distance from the node calculated to give the position in the first-anomaly model, relative to the node.

If all of this sounds excessively subtle, that is perhaps because it *was* excessively subtle, even for Tycho's observations. For Mars the greatest correction that should ever be applied in a reduction to the ecliptic is less than 53 seconds of arc, at positions 45° from the nodes. Tycho's assistants, in addition to calculating erroneous times for the acronychal observations, had been applying corrections as large as 9 minutes. This was partly due, as we shall see, to the pervasive confusion in pre-Keplerian latitude theory; but it must have been partly due to their own lack of competence.

It seemed evident to Kepler that, since Mars traveled around the sun and was moved by it, the orbit should lie in a plane with the sun. Current theories denied this. Ptolemy had supposed, naturally enough, that the planetary eccentrics intersected at the center of the universe, the earth. In his detailed latitude theory, he had tilted the epicycles back parallel to the ecliptic. However, he had been forced to rock them back and forth to make the theory work. The reason for this, Kepler argued, was that the orbital planes actually intersect at a point, the sun, which does not appear in those planetary models.[22] Copernicus, one would think, should have cleared up the disarray.

[21] Kepler called them *tabulae conditores,* and did not remark on whether he thought them to have acted on Tycho's instructions or their own initiative. *G. W.,* 3: 115: 17–21.

[22] For Ptolemaic latitude theory see O. Pedersen, *A Survey of the Almagest* (N.p.: Odense University Press, 1974), pp. 355–386; O. Neugebauer, *A History of Ancient Mathematical Astronomy* (New York: Springer-Verlag, 1975), 1: 206–230; R. C. Riddell, "The Latitudes of Mercury and Venus in the *Almagest,*" *Archive for History of Exact Sciences* 19 (1978): 95–111.

As we have remarked, though, the sun itself did not appear in Copernican planetary models either. As a result, his latitude theories remained complicated, with synodic components that were all the more surprising after the heliocentric transformation had eliminated the synodic epicycles of longitude theory.[23]

Kepler, of course, was right in thinking that his own elementary physical considerations could simplify all this. Planetary orbits lie in planes passing through the sun. The orbit of Mars is inclined to the ecliptic at an angle of 1;50°. Kepler found that Tycho's assistants, perhaps inured to bizarre theories of latitude, had been using one with a "fractured" eccentric, wherein the maximum northern latitude of Mars was 4;33°, but the maximum southern latitude was 6;26°.[24] Part of the problem, it seemed, was that they were not properly taking into account the rather obvious fact that observed latitudes are increased when the earth is close to the planet, and decreased when it is distant. Even this large optical effect, which Kepler aptly termed the second inequality of latitude, did not account for all of the problems he found in the reduction of Tycho's observations. They had been badly handled and needed a thorough reworking. Latitude theory had to be completely redone. Surely, Kepler thought, the sun was moving the planets in simple planes passing through its body. Only after getting this straight could he go back to the raw observations, to determine the moments of true opposition and the longitudes of the planet at those moments.

Kepler required six chapters, 10 through 15, to describe and justify his reduction of the Tychonic observations. He demonstrated that the diurnal parallax of Mars was negligible; he successfully established his simple latitude theory; and finally he adjusted the acronychal observations to the moments of true opposition. Although he had already devoted some effort to arguing that true oppositions should be used instead of mean oppositions, he postponed this change until the other preliminaries were completed so that no one would object that he was "hiding behind the thickets of his own method."[25] To appreciate properly Kepler's accomplishments in the *Astronomia nova* one really should not neglect these chapters, so important in realizing the precision of Tycho's observations; but they would lead us far from our topic.[26] We shall only remark briefly on his latitude theory, for this was both physically motivated and of great importance to the development of astronomy.

As explained above, Copernicus's latitude theory had not achieved the simplicity inherent in the heliocentric viewpoint, because it had been centered on the mean sun rather than the sun itself. He had been "ignorant of his own

---

[23] N. Swerdlow, "The Derivation and First Draft of Copernicus's Planetary Theory: A Translation of the Commentariolus with Commentary," *Proceedings of the American Philosophical Society* 117 (1973): 484–489.

[24] *G. W.*, 3: 116–117.

[25] *G. W.*, 3: 130: 5–8.

[26] See R. Small, *An Account of the Astronomical Discoveries of Kepler* (originally published London, 1804; reprint Madison: University of Wisconsin Press, 1963), pp. 163–177, 324–330.

riches"; but Kepler, "armed with his incredulity,"[27] had no hesitation in jettisoning all the elaborate apparatus by which Copernicus had reproduced the appearances of Ptolemaic latitude theory. The simplification Copernicus achieved in replacing the separate but coordinated second-anomaly models of longitude by a motion of the observer was incomplete, Kepler protested, so long as the system required *any* motions tied to, but not explained by, the earth's motion. Whatever it was that moved Mars should not be obliged to notice the position of another planet.[28]

This was not, of course, merely a physical argument. Kepler was appealing as broadly as possible to the simplicity and explanatory power of helio-centrism, and these advantages can be variously conceived as physical, logi-cal, aesthetic, or simply pragmatic. An explicitly physical discussion of motion in latitude was postponed until the fifth and final part of the book (below, pp. 130–137). Nevertheless, Kepler wasted no time in stripping latitude theory to the essentials: the sun, the planet, and a plane containing the planet's orbit. This simple model proved consistent with observations taken from various points in the earth's orbit, which confirmed Kepler's opinion that the plane-tary motions were physically independent.

With all these preliminaries on record, Kepler in Chapter 16 finally set out to construct a planetary model. What he had in mind was an equant model, "in imitation of the ancients" as he said, but without Ptolemy's restriction that the the eccentricity be precisely bisected. Derivation of the parameters for such a model required four acronychal observations (Figure 8). The orbit was centered on point B, which was distant from the sun at A by an eccentricity to be determined. The equant center C was likewise eccentric by an amount to be found empirically. As we shall see, the placement of C on a line with A and B was not empirical, but was a physical constraint Kepler placed on the model.

To specify this model completely one needs four parameters: the two eccentricities AB and BC, in proportion to the radius of the circular orbit; the direction of the apsidal line IH, on which A, B, and C lie; and the mean anomaly of one of the observations.[29] The two eccentricities are part of the geometry of the model. The apsidal longitude and mean anomaly locate the orbit spatially and temporally, or, what comes to the same thing, they establish a connection between the observed longitudes and times and the model. In order to investigate the eccentricities, Kepler had to connect the observations to the model. That is, he had to assume values for the longitude of aphelion and for the mean anomaly of an observation. From the known angles between the observations, and the corresponding time intervals, Kepler had the angles between the planetary positions, as measured both at the sun A and at the equant center C. He therefore knew the four triangles with vertices D, E, F,

---

[27] *G. W.*, 3: 141: 3–16.

[28] *G. W.*, 3: 131: 1–6.

[29] The mean anomaly of an observation was simply the time, as a proportion of the planet's period, between aphelion and that observation. If it was known for any one observation, the intervals between observations gave it for all others.

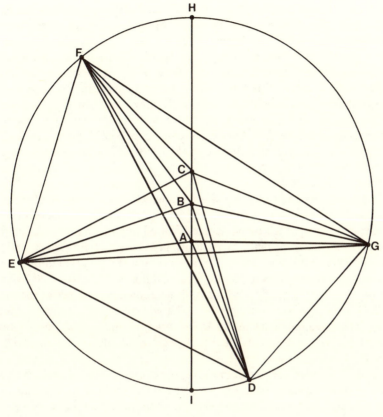

Fig. 8

and G (the modeled planetary positions) from AC as a base; and in particular
he knew the *distances* from the sun to the four positions implied by the model.
These distances depended upon the assumed direction of the apsidal line,
the assumed mean anomaly of one of the observations, the observed angles
around A, and the observed intervals of time, which are the angles around C.

The model was further restricted, for Kepler had not yet introduced into
his calculations the assumption of a circular orbit. Four pairs of radii from A
and C certainly intersected at four points; but were these four points on a
circle? Any three (non-collinear) points lie on a circle, but in general four do
not. If not, he had to change one of the auxiliary assumptions, the direction
of the apsidal line. When eventually an apsidal line yielded four points that
satisfied the geometrical requirement, lying on a circle, Kepler introduced a
further requirement *a priori*: that the center of the circle lie directly between
the sun and the equant center.

Kepler could have justified this requirement in any number of ways. After
all, no one had ever proposed that the center of a planet's orbit might lie

elsewhere than on its apsidal line. Kepler's explanation, however, was based on his physical interpretation of equant motion: "physical reasons would require that the motion be slowest where the planet is most distant from the sun A, as in H; which cannot happen unless A, B, C are in the same line." [30]

Imposing this physical assumption on the model squeezed out the last bit of freedom in choice of parameters. As before he had tried various apsidal lines to make the four points lie on a circle, now Kepler tried various values for the mean anomaly to bring the center of the circle directly between the sun and the equant center. After each revision, of course, he had to readjust the apsidal line. Kepler said that he worked through the calculations seventy times before arriving at his final parameters for an equant model with non-bisected eccentricity.[31]

The outcome of all this labor was a model wherein the center of the circle was eccentric to the sun by 0.11332 of the radius, and the equant center was eccentric to the sun by 0.18564. Kepler showed that this model, derived from four of Tycho's acronychal observations, predicted all ten of them, and two more observed by Kepler himself, to an accuracy of about two minutes. This was as accurate as he believed the observations to be, and indeed was no greater than his (erroneous) estimate of the apparent diameter of the planet itself. He concluded that since his theory based on the true sun was as accurate as Tycho's based on the mean sun, the accuracy of the latter was no argument against his insistence on using the physical body of the sun in planetary theory.[32]

Thus far Kepler's physically-motivated reforms had fared well. If he was right in building his astronomy around the sun, he had shown that other astronomers could have built models around the mean sun that were accurate in predicting longitude, though not in predicting distances. If he was right, motion in latitude followed orbital planes through the sun and of constant inclination, not planes through the mean sun afflicted with inexplicable tiltings and rockings. Moreover, he had actually constructed a longitude model as good as even that of Tycho, but with the further advantage that the planet's variation in speed was tied to its distance from the sun. Had he stopped there, after eighteen chapters, Kepler would already have contributed much to the refinement of Copernican astronomy. Instead, he immediately demonstrated in two different ways that his own theory remained inadequate. To be sure,

---

[30] *G. W.*, 3: 155: 13–15.

[31] For detailed exposition of the procedure, see Delambre, 1: 409–417; Small, pp. 180–185. O. Gingerich has shown that Kepler made so many calculations primarily because he started over repeatedly, with different sets of observations, between 1600 and 1604. The nested iterations were not as laborious as they first appear, because previous theory provided good initial values for the apsidal line and mean anomaly, and because the results of each adjustment usually indicated the direction and approximate size of the next adjustment. The calculations must still have been very tedious. "Kepler's Treatment of Redundant Observations," *Internationales Kepler-Symposium, Weil-der-Stadt 1971*, ed. F. Krafft, K. Meyer, B. Stickler (Hildesheim: Verlag Dr. H. A. Gerstenberg, 1973), pp. 307–314.

[32] *G. W.*, 3: 174: 9–12.

it performed the function of a theory of longitude. The twelve acronychal observations were scattered all around the zodiac, and the theory gave the right longitude for all of them. What it did not give was the right location for the planet itself.

Sensitized, perhaps, by his earlier analysis to the possibility that two quite different models could predict virtually the same longitudes, Kepler devised two methods of checking whether Mars was as far distant from the sun as his model implied. The first method involved geocentric latitudes. These depended, of course, upon both the heliocentric latitudes and the earth's position relative to the planet. Kepler had already shown, by methods relatively insensitive to the distances of the earth and Mars from the sun,[33] that heliocentric latitudes were determined by the orbital plane's constant inclination of 1;50° to the ecliptic. At opposition, geocentric latitudes much greater than this had been observed, since the earth was then only about one-third as far from Mars as was the sun. The two latitudes sufficed to determine a triangle from which the ratio of the distances of the earth and Mars to the sun could be estimated. Little precision was possible, since the angles involved were so small that any error in them had a relatively great impact.

Kepler solved such triangles for the oppositions of 1585 and 1593, when Mars was respectively near aphelion and perihelion. Correcting these distances to their values at precisely the apsides, and comparing them, he found that the eccentricity of the orbital center was in the range .08 to .10. His model of Chapter 16 had put this eccentricity at .113 + (p. 44, above). He now knew that the earth was too close to Mars at the latter's aphelion, and not close enough at perihelion, for this value to be correct. Kepler further noted that, curiously enough, the Ptolemaic principle of a bisected eccentricity was consistent with the geocentric latitudes at opposition. The eccentricity of the equant was .18564 in his model, and if the orbit was centered halfway between the sun and the equant center it would have an eccentricity just about what the latitudes indicated.

It was not accidental that analysis of the distances between Mars and the sun led Kepler back to the hypothesis of bisected eccentricity. The area law, which governs the motion of a planet on its ellipse, can be well represented by equant motion around the empty focus of the ellipse. Thus the center of the ellipse bisects the eccentricity of its pseudo-equant point at the empty

---

[33] We are oversimplifying here, where the issues are not physical. Kepler had derived the inclination of the orbit by three methods in Chapter 13; they are described in Small, pp. 167–174. Of these the third proceeded from geocentric latitudes and from the ratio, given by previous theory, of distances at opposition; it was thus the inverse of the argument of Chap. 19, summarized in the text. We should hardly be surprised, then, that Chap. 19 confirmed the ratio of distances given by previous theory, that is, by a bisected eccentricity. The first method likewise depended upon approximate knowledge of the distances. The second method, however, was independent of the distances. When the earth is in the line of Mars's nodes and the elongation of Mars from the sun is 90°, the geocentric latitude of Mars equals the inclination of its orbit. Historically, Kepler's work on the non-bisected eccentricity and his work on latitude theory may have been very tangled; see Gingerich, p. 312.

focus. (This is why Ptolemy's first-anomaly models were so good.) Moreover, the actual ellipses are virtually indistinguishable, in shape, from circles. A circular orbit with bisected eccentricity therefore represents the actual distances reasonably well.

It does not represent the longitudes as well as one with non-bisected eccentricity like that of Chapter 16. That model had predicted longitudes within about two minutes. With the eccentricity divided equally, the error grew to about eight minutes near the octants. Even this, Kepler pointed out, had been an acceptable level of error prior to Tycho's refinements in observational technique: Ptolemy would have had no reason to suspect such a model's accuracy in representing acronychal observations. Kepler, knowing the accuracy of Tycho's observations, could not neglect error of this magnitude. The original model of Chapter 16 had failed to represent distances, but the model corrected to do that failed to represent longitudes.

Knowledge of Mars's distance from the sun was needed not only to predict geocentric latitudes, but also geocentric longitudes. Kepler chose a pair of observations when Mars was at aphelion and perihelion, and the earth somewhat off to the side. Based on the known heliocentric longitudes of both planets, and rough values for the earth's distance from the sun at those times, he again showed that the eccentricity of the Martian orbit was about .09, not .11 as indicated by his model derived from acronychal observations. His theory of the first anomaly of Mars was a failure, even though his free division of the eccentricity made it more general than prior models. He had done the best he could within the traditional framework, a circular path with the planet moving uniformly around an equant center. Henceforth he would have to try something radically new.

Two features stand out from Chapter 20, where Kepler for a second time refuted his own model. The first is that, although his equant model had failed him, it did at least give very accurate heliocentric longitudes, within about two minutes of those observed. These were not all that he wanted, but they were quite a lot nonetheless. Whatever problems awaited him still, he could always use this imperfect theory to locate the planet on a line from the sun. In terms of the traditional astronomer's task of saving the phenomena, he needed only a theory of distances. Already he had used the longitudes from this theory of Chapter 16 in one of his refutations of that very theory. Because it was so useful and yet wrong, he called it a "vicarious hypothesis," and made it serve as a stand-in, doing some of his work for him while he sought the true theory.

The second feature revealed by Chapter 20 was that once the traditional practice failed, of modeling the first anomaly from acronychal observations, one could no longer postpone the question of the earth's own location. Outside of opposition to the sun, the second Martian anomaly had to be taken into account, and this depended upon the position of the earth. In particular, accurate knowledge of the earth's distance from the sun at various points in its orbit was essential to any precise use of the observational data. The whole question of distances could not be broached without a scale for comparison, and Kepler could not hold this scale constant unless he knew the variations in the distance between the earth, where Tycho had observed, and the sun, to which Mars's motion had to be referred.

Before leaving the realm of traditional astronomy, Kepler paused for a brief incursion into philosophical ground, to address the question of how a false hypothesis had been able to produce true results. His discussion, however, was mathematical and was specifically focused on the question of modeling a planet's motion. Today it seems a little odd. His vicarious hypothesis was wrong, but produced longitudes accurate to the limits of his observational data. Was this not an instance of a false hypothesis yielding truth? Kepler denied this for two reasons: first, it was correct only in longitude, and not in distance. Furthermore, its longitudes were not necessarily true; merely accurate within the limits of perception.

This was the central idea of the chapter. Innumerable theories were possible, even with the restrictions of a circular orbit and motion that was regular around an equant center. Kepler showed, by example, how one could save the phenomena at the apsides, quadrants, and octants successively, by choosing among relatively simple models. For Mars, a concentric circle accounted for the fact that passage between aphelion and perihelion always required the same amount of time, half the period of revolution. An eccentric would place the planet correctly at each quarter of the periodic time, still erring at the octants, but by less than half a degree. By separating the center of the orbit from the equant center one could reduce the errors at the octants below two minutes. At each step the error remaining in intermediate places decreased markedly. Clearly it would be possible to reduce the error below perceptible limits, by making a relatively small number of *ad hoc* adjustments, but without having any reason whatsoever to think that one had reached the truth at the moment one stopped noticing the error. A theory could be accurate within observational error and yet be wrong, both physically and numerically, even though no numerical error was detected.[34]

Where a false theory seemed to yield true results, then, Kepler raised two cautions. First, the conclusions might seem right only because of the dullness of our perception. Second and more generally, a true result only reflected credit upon those parts of the hypothesis that were responsible for the particular truth of that result. Unless one understood exactly how the result followed from the hypothesis, one could not assess the degree or extent of empirical corroboration.

One motive behind this chapter, which digressed from the momentum of the book's argument, was surely that given explicitly by Kepler: "... I vehemently hate this axiom of the dialecticians, that truth follows from falsehood,

---

[34] Kepler expressed this caution clearly, using the same series of astronomical examples as in this passage, in the long letter he finally sent to Fabricius in July of 1603, letter #262 in *G. W.*, 14: 425–426.

because by it the throat of Copernicus is attacked (which teacher I follow in
the more universal hypotheses of the world system) .... "[35] All his life Kepler
insisted on a literal interpretation of the Copernican hypothesis, that the sun
really was the immobile center of the planetary system. He had himself
published, on the back of the title page to the *Astronomia nova*, the fact that
Osiander and not Copernicus had written that unsigned preface to *De revolutionibus* which characterized heliocentrism as a computational device without
claim to reality. If astronomers could compute the same equations from
heliocentric and geocentric models, that did not mean they should relinquish
the choice between world hypotheses to logicians and theologians. Their
proper conclusion was merely that special characteristics of the Copernican
and Ptolemaic theories, the geometrical models for the first and second
anomalies, produced results that were the same in certain senses, namely the
equations and longitudes. Only in this limited way was it true that the two
hypotheses were astronomically equivalent.

On the surface, then, Chapter 21 made the point that it was a relatively
trivial matter for two hypotheses to agree approximately in some of their
predictions, even very important predictions. One cannot help remarking that
at this level Kepler's argument ignored some salient points, which we may
summarize as follows. Mathematical astronomy was an ancient science, and
its goal was to save the phenomena, that is, to predict as accurately as possible
the places and times of what could be seen in the heavens. Insofar as two
hypotheses were equally successful at this goal, the distinction between them
lay outside mathematical astronomy. One could choose only by appealing
outside of astronomy. Kepler had succeeded in isolating a phenomenon, the
distance, which was distinct from the equations of longitude and latitude, but
he had done this only by assuming a particular Copernican theory which he
ended by disproving. Either world hypothesis could be arranged to save the
phenomena of distance, and if one hypothesis did so as well as the other,
astronomy could not choose.

By traditional standards, this would have been a fair defense of mathematical astronomy from Kepler's attack, and would have explained why an
astronomer might fairly have rejected that attack. We will not pause here to
distinguish between Copernican distances, which follow necessarily from the
basic assumptions of the system, and geocentric distances, which one might
adjust to the results of subtle analysis like that of Kepler; because it is not our
purpose to establish the superiority of Copernican cosmology. Instead we
merely observe that, in historical context, the above objections were eminently
reasonable. Models for the first and second anomalies may have been special
characteristics, equivalent as predictors of longitude, within the more general
cosmology; but of mathematical astronomy they were the essence. It was in
this sense one said that astronomy could not choose between Ptolemy and
Copernicus. Kepler was at a critical point in the development of a new

---

[35] *G. W.*, 3: 183: 4–6.

astronomy. Although he did not say so, he really was appealing outside of mathematical astronomy, the science of saving the phenomena. He appealed, however, not to theology or dialectic, nor even to the empirical physics of his time, but to his own conception of astronomy as at once physical and mathematical. This conception was the finest fruit of Copernicanism—which itself had sprung, ironically, from an attempt to accomodate astronomy to other and very different physical principles. Kepler's work was a crucial step in the creation of the physical science that astronomy is today.

However obvious these things are in retrospect, the need to reformulate the principles, the methods, and the very goals of his science must have given pause even to the confident young Kepler. From this perspective, we can see the twenty-first chapter as something far more interesting than a half-articulated defense of the realist interpretation of Copernicus. It was a synopsis, so brief as to be a caricature, of the history of astronomy, as it appeared to the first man who was in a position to appreciate its futility. If the task of geometrically reproducing the appearances had never before seemed so arbitrary and pointless, that was because there had always been elegant assumptions, of circularity and of uniform motion, which had ordered the astronomer's task, guiding him through the infinite possibilities of geometry; and which at the same time had suggested the mode of their own realization by regularly moving spheres. Tycho had removed this latter prop, and Kepler's analysis of Tycho's observations indicated something amiss with the assumptions. Another adjustment was due in the geometry; but more than that, Kepler could finally see that just another adjustment would not be enough. If he was ever to understand why a geometrical model did or did not work, he had to find, and learn to calculate, how the planet moved or was moved along its orbit. He concluded the second part of his work, devoted as he said to imitation of previous workers, and inaugurated his new study of the heavens by commencing a study of the earth.

Theory of the Earth

The theory of the earth, or solar theory as it is conveniently known, was in need of some reworking. Essential to the first development of lunar and planetary theory, solar theory had lagged behind them in the work of Ptolemy and hence of later astronomers. The reasons are evident. The sun is not seen against a background of stars by which its position may be readily known.[36] One knows it to be precisely opposite the moon at lunar eclipses, but because of the moon's rapid and complex motion solar theory must precede lunar theory.

As a result, the solar theory in use at the beginning of the seventeenth

---

[36] Furthermore, the stars themselves were mapped with the aid of lunar theory, which in turn depended on a rough solar theory. See the discussion in Neugebauer, 1: 53ff.

century was both ancient and primitive. Hipparchus had suggested a simple model for the sun, a uniformly rotating eccentric circle, and had shown how to derive the parameters of this model from consideration of the lengths of the seasons. Ptolemy (*Almagest*, Book three, Chapter 4) had retained this theory unchanged, without separating the center of the circle from the center of equal motion as he had for the other planets. His successors also had retained both the fundamental assumption of an eccentric circle rotating uniformly about its center, and Hipparchus's method of deriving the parameters.[37]

In its guise as the second anomaly of the planets, solar theory was still more primitive. The epicycle of the second anomaly rotated uniformly about its center, the same point which attached it to the first-anomaly model. In hindsight, we can see that this model corresponded to the simplest imaginable solar theory, a concentric circle rotating uniformly. It did not mean this to a geocentric astronomer, of course, and so Copernicus was the first to have to take the equivalence seriously. He duly modeled the second anomaly with the simplest imaginable solar theory—now of course a theory of the earth—a concentric circle rotating uniformly. This meant that, as we have already noted, his models for the other planets were attached not to the sun, but to the center of the earth's circle, the "mean sun." These two facts neatly preserved the equations of the Ptolemaic second-anomaly model.

That second-anomaly model, and the theory of the earth in which it was now imbedded, were old and crude, and needed some critical examination. Already Tycho, stimulated no doubt by his own theory of the second anomaly, had determined that all was not well. The uniformly moving concentric, he had written to Kepler in 1598, appeared to expand and contract.[38] We can only assume that this was a hypothesis worthy of consideration to Tycho, imbued with an astronomy of circles and uniform rotation. (It may have seemed reasonable to Kepler too in 1598; his marginal annotations to that letter concerned Tycho's reception of the *Mysterium* and his objections to Copernicus, but not the fluctuating size of the mean sun's circle.) At some point in the intervening years Kepler had realized that Tycho's passing remark held the key to one of the doubts he had expressed, at the end of his *Mysterium*, regarding his physical explanation for the changing velocities of the planets. If the other planets accelerated when they approached the sun, and decelerated when they withdrew, why did the earth on its eccentric circle always move with the same speed?

The fluctuating size of the earth's orbit solved this puzzle. Considered simply as a comment about the motion of the body of the earth, Tycho's statement was that the earth was not always at the same distance from the mean sun, the center around which its angular motion was uniform. The

---

[37] For example, Copernicus, Book three, Chap. 16.

[38] *G. W.*, 3: 191: 26–32; quoting from letter No. 92, *G. W.*, 13: 198: 51–55. In the following discussion we shall continue to neglect the Ptolemaic and Tychonic interpretations of the second anomaly. Kepler, however, thought the analysis of Chaps. 22–26 sufficiently important that he presented it in all three interpretations.

earth's orbit was eccentric to its equant, just like those of the other planets. Doubtless (Kepler thought) the orbit was centered between the sun and the equant center, so that the earth also varied its speed with its distance from the sun.

Kepler proceeded, as best he could, to the measurement of the earth's orbit. This was no easy job. He could judge the earth's position only by comparing the observed longitudes of Mars to the heliocentric longitudes, with the latter being derived from existing theory. But existing theory, although it gave good heliocentric longitudes—as could be verified at opposition—did not tell him the distance of Mars from the sun. Since Kepler did not know where Mars was, he simply avoided that question by using groups of observations separated by 687 days, the period of the planet's revolution. Wherever it was, it was at the same place each time. This procedure allowed him, in effect, to observe the earth from a stationary Mars, by reversing the directions along which Tycho had observed Mars from the earth.

One notices again that this method was entirely dependent upon Tycho's passion for observing the planets in all different configurations. Previous astronomers had been clever, observing chiefly at critical moments, such as oppositions to the sun, when their observations gave clear answers to the questions they wanted to ask. Tycho's less directed program proved its worth, naturally enough, when Kepler's research reached the point where he had to ask questions that had not been thought of before.

Before plunging into the analysis, Kepler modestly consented to use, for the time being, the traditional assumption which tied the planetary motions to the mean motion of the sun, "lest someone think my innovation, in using the apparent motion of the sun, is implicated in this matter." [39] He went back to the mean sun as the connecting point for the apsidal lines, a hypothesis which he had assailed as physically absurd. His ostensible reason was to avoid controversy in making an important point about the geometry of the earth's orbit; a good and adequate reason. Notice, though, what he wanted to establish: that the center of the earth's uniform angular motion was different from the center of its orbit. Mars's longitude as measured from the sun itself would have been of no direct use, for he was not trying to prove that the sun was eccentric in the earth's orbit. That, or its equivalent, had been long known. He wanted to show that the earth's equant center, the mean sun, was eccentric in the orbit; and to do that he needed to connect his effective observation post (on Mars) with the mean sun. He therefore used Tycho's theory of Mars. Like the vicarious hypothesis, Tycho's Martian model gave bad distances but good longitudes. Its physical implausibility was irrelevant. With it, from Mars, he could locate the mean sun at any time; and from Mars he could locate the earth with the observations Tycho had made of Mars. Kepler was not wavering in his physical convictions. If his readers had more confidence in results obtained using Tycho's conventional theory, Kepler was willing to accept

---

[39] *G. W.*, 3: 192: 19–21.

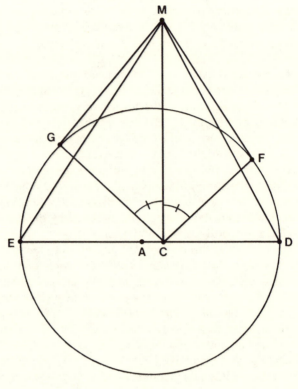

Fig. 9

their confidence; but that was not why he used the Tychonic model here. He
simply took the most direct route to his conclusion, a conclusion which after
all confirmed his physical point of view.

His first method, presented in Chapters 22 and 23, led directly to the result
he wanted. As always, the cost of choosing such a convenient method lay in
the stringent conditions upon observations suitable to it. In Figure 9, C is the
mean sun and A the true sun. As remarked above, the true sun has nothing
to do with Kepler's intended conclusion, that the orbit is eccentric to C. It
did, however, tell him in which direction to expect the eccentricity. He presumed that the earth, like the other planets, actually moved slowest when
farthest from the sun. Now, the velocity modeled by an equant is least where
the planet is closest to the equant center. Physically, therefore, the center of
the circle belonged not at C, but directly between A and C, so that the planet
would be closest to the equant center when farthest from the sun. But AC was
the earth's apsidal line, known to lie approximately from Cancer $5\frac{1}{2}°$ to
Capricorn $5\frac{1}{2}°$. If Kepler could find a pair of observations in which the earth

occupied either apse (points D and E), while Mars was each time at the same point M, at right angles to the apsidal line DE as seen from C, he could immediately tell whether C was the center of the earth's orbit. If C was the center, angles CMD and CME would be equal. If, as he suspected, the center lay more toward A, angle CME would be greater than CMD.

Such precisely specified observations were not to be found. Indeed, Kepler pointed out that the periods of Mars and the earth were incommensurable, so that no small number of integral revolutions of Mars would find the earth at places separated by exactly a semicircle. He therefore relaxed his search by requiring only what was essential in Figure 9. Mars clearly had to be out to the side of the earth's apsidal line, around Aries or Libra, so that the distances on either side would be perceptibly different. The two positions of the earth did not have to be right at the apsides; but if a simple comparison of angles was to suffice, it was essential that the two positions of the earth be distant by equal angles from the line CM connecting Mars and the mean sun. That is, they had to be position such as F and G in the figure, where angles MCF and MCG are equal. (Such angles, called *angles of commutation* by Copernicus and Kepler, sufficed along with the relative distance to Mars to compute the equation of the second anomaly in Copernican theory, since Copernicus assumed C to be the center of the earth's orbit. Kepler was arguing the opposite in his critique, that because C was not the center, equal angles of commutation could produce different equations, even with Mars at exactly the same distance.)

This sort of observation-pair could be found in Tycho's books. As always, the recorded observations were not at precisely the right moments; but over a day or two Kepler could accurately adjust them, preferably by interpolation but when necessary by using the planet's velocity as known (approximately) from conventional ephemerides. As he had expected, the equation or angle CMG turned out to be more than a degree greater than CMF. Exploring the configuration in detail trigonometrically, he calculated that the center of the earth's orbit was distant from the mean sun by about .018 of the radius, in the direction of the sun itself. Tycho's solar theory gave the distance between the mean sun and the true as .03584, so the earth's orbit seemed to agree with all the others in being centered halfway between the solar body and its own equant. This conclusion, if reliable, implied that the earth was moved like the other planets.

An assertion so broad in its implications could not be left in doubt. Kepler began anew in Chapter 24. He retained the device of spacing observations by the 687-day period of Mars, so that he could pretend to observe the earth from a fixed position of Mars; but he no longer insisted upon particular situations of the earth. With four observations he laid out a configuration like that of Figure 10. Again C is the mean sun; M is Mars, whose longitude from C was known by Tycho's theory; and the four points E represent the earth at the moments of observation. Their longitudes from C were known because C,

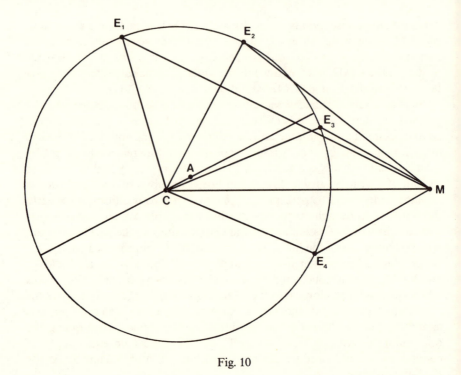

Fig. 10

by definition, was the center of uniform angular motion for the earth.[40] The
remaining lines in the figure, connecting M with the points E, were simply the
observed longitudes of Mars. For each observation, Kepler therefore had all
of the angles in the triangle earth-Mars-mean sun; so by the law of sines he
expressed the four distances between earth and mean sun in terms of the
common distance MC. Again the distances were unequal, and again they
indicated that the terrestrial orbit was closest to its equant around aphelion—
that is, that the earth moved most slowly when it was farthest from the
sun.

The four distances determined a circle, or rather four nearly-coincident
circles, depending upon which three were used. Numerical results were not so
good as with the former method, since this configuration was not especially
suited to estimating the parameters accurately. Nevertheless his conclusion
stood: "the center of the earth's revolution is at an intermediate place between
the body of the sun and the equant point of that revolution: that is, the earth
moves nonuniformly in its orbit; it becomes slow where it withdraws from the

---

[40] Actually, this fact and the earth's periodic time give only the angles between the points E. For
actual longitudes Kepler had to assume an epoch, and hence an apsidal line. Again we see how
important it was that Kepler was willing (happy!) to suppose that the earth's apsidal line passed
through the sun, and thus in the direction where Tycho had put it, *mutatis mutandis*.

sun, swift where it approaches; which is agreeable to physical considerations and to the analogy of the other planets."[41]

Having dislodged the earth's orbit from the mean sun, Kepler in Chapter 26 investigated its disposition with regard to the true sun. He used the same method as before (Figure 10), but with central angles taken at the true sun. His vicarious hypothesis provided the true heliocentric longitude of Mars. Positions of the earth with respect to the sun came from Tycho's solar theory, which was accurate enough regarding longitudes. (Later, in Chapter 31, Kepler confirmed that Tycho's failure to bisect the solar eccentricity did not sensibly affect the longitudes of his solar theory.) The solution was approximately what Kepler wanted, that is, he found the earth's positions to lie on a circle only about half as eccentric to the sun as formerly thought. He worried the observations for several pages, showing how small, even imperceptible changes could appreciably alter the computed eccentricity and apsidal line. As he pointed out, though, the place of the solar apsidal line was known by other and less sensitive calculations; and his minute adjustments, while bringing the apsidal line closer to its true place, also brought the eccentricity closer to .018, or half its previous value.

Thus far in the book, Kepler had generally sketched out the equivalent Ptolemaic and Tychonic versions of his demonstrations relating to the second anomaly. In Chapter 26 he did so once more, at some length. All of his work on solar theory, in Book III, had been based not on observations of the sun but on observations of Mars, controlling the first anomaly through separation of the observations by integral revolutions of that planet. From the viewpoint of purely mathematical astronomy, then, he had been working not necessarily on the theory of the earth or the sun, but rather on the theory of the second anomaly of Mars. In Copernican cosmology, this was necessarily the theory of the earth's motion; in Tychonic cosmology, of the sun's motion. In Ptolemaic cosmology it was the theory of the epicycle. Supposing that one accepted Tycho's observations and Kepler's mathematics as sound, one might indeed feel compelled to transfer the results over to solar theory. A skeptical or contrary reader might not. At any rate, the necessary geometrical equivalent of his conclusions, in Ptolemaic astronomy, was to be sought in the epicycle saving the second anomaly of Mars.

The consequences of this equivalence were devastating. One had to speak of the epicycle's "point of attachment" (*punctum affixionis*—G. W., 3: 211: 34), rather than simply its center, because the epicycle was no longer attached to the eccentric by its center. Instead, a point near the center of the epicycle (and corresponding to the sun if one viewed the epicycle as a transformed image of the earth's orbit) was attached to the eccentric first-anomaly model. Moreover, the motion of the planet around the epicycle became physically nonuniform, because the planet moved uniformly around still another point, corresponding to the earth's equant.

---

[41] *G. W.*, 3: 203: 29–33.

This meant that the epicycle's center traced an eccentric of its own, parallel to the eccentric of the first anomaly, but whose apsidal line did not pass through the earth at all. The planet's velocity on the epicycle depended on how it was positioned with respect to the point of attachment and the zodiacal location of the solar (!) apsidal line. Other rearrangements of the epicycle were possible, but all were hopelessly complicated. One ended by attaching to each planet a full-blown model of Kepler's new solar theory, complete with bisected eccentricity.

There is irony here. Discarding Copernicus's inadequate second-anomaly model in favor of one more elaborate, Kepler had ended by demonstrating that heliocentric theories of the second anomaly were necessarily simpler than geocentric theories. He predicted that "the Sun will melt all this Ptolemaic apparatus like butter, and the followers of Ptolemy will disperse partly into the camp of Copernicus, partly into that of Brahe." [42] Henceforth he did not discuss the Ptolemaic theory as a serious possibility.

Before proceeding with his analysis of the earth's motion, Kepler devoted two more chapters to refining and cross-checking the parameters he had obtained against further observations, to reassure himself of their consistency. In retrospect this sort of redundant calculation appears digressive and unnecessary, but it was his only way of testing his assumptions. Kepler's understanding of the interplay between observation and theory was most delicate, one of the characteristic features of his genius.

At this point, Kepler had successfully relocated the earth's orbit, which was now centered at half its former eccentricity from the sun. He had gone to this trouble in order to be able to use observations taken from any point in the earth's orbit. Therefore he constructed, and laid out in Chapter 30, a table of the earth's distances from the sun for every degree of a semicircle of anomaly. The table is exceptionally confusing, because it was not published in its original form. By the time Kepler published it he knew enough to recast it into a form that would make it still useful at a later point in the book, after he had disproved the theory on which it was based.

So far as the reader was concerned in Chapter 30, the table simply gave the distances between the sun and earth, for every degree of true anomaly from 0° to 180°. It was constructed on the hypothesis of a circular orbit, and normed to a radius of 100,000 and an eccentricity of 1800. The odd thing was that the integral angles of true or coequated anomaly, upon which the construction and use of the table were based, did not appear in it. Instead, two columns of angles accompanied the distances in the table. One was labeled "coequated anomaly," but Kepler explained that this column really contained angles equal to the coequated anomaly minus the optical equation, angles which therefore had no meaningful interpretation. The other column, labeled "mean anomaly," contained not the mean but the eccentric anomaly, that is, the coequated anomaly plus the optical equation. To be sure, the coequated anomaly was

---

[42] G. W., 3: 213: 19–21.

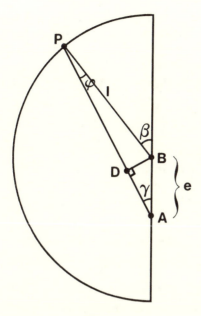

Fig. 11

readily apparent, as it was an integer squarely in the middle of the two given angles, which themselves differed at most by a couple of degrees.

As Kepler further explained, one could safely take the tabulated angles to be what they were labeled. This would give distances too short to reach the circle; distances, hence, not to the circle but to an oval within it. The actual distances, he hinted, would turn out to be short in just this way—but only by half so much. In reality, the whole table is a remarkable compromise between the theory Kepler had in Chapter 30 and his final theory. It was constructed to agree precisely with his earlier theory, but laid out so that its natural interpretation lay close enough to both theories to be usable for either. Let us pause to see these relations, since they have evidently been explained nowhere else.

For a true or coequated anomaly γ, the distance AP to the planet (Figure 11) is given by

$$AP = AD + DP$$

$$AP = e \cdot \cos\gamma + \sqrt{(1 - e^2 \cdot \sin^2\gamma)} \tag{8}$$

Since $e = .018$, $e^2 \cdot \sin^2\gamma$ is very small and we may approximate

$$AP = e \cdot \cos\gamma + 1 - (e^2 \cdot \sin^2\gamma)/2$$

$$AP = 1 + e \cdot \cos\gamma - (e^2 \cdot \sin^2\gamma)/2 \tag{9}$$

Equation (8) corresponds to Kepler's method for calculating the distances, as given in Chapter 29. The distances in the table are thus derived from (8), for an angle γ midway between the tabulated angles. Equation (9) is equivalent, to the 5-place accuracy of the tables.

The optical equation, angle BPA in Figure 11, is simply

$$\phi = \sin^{-1}(e \sin \gamma) \tag{10}$$

$$\phi \approx e \sin \gamma, \quad \text{for } e = .018$$

The tabulated angles are just $\gamma \pm \phi$.

These angles, however, are labeled "mean anomaly" and "coequated anomaly," which if taken literally would imply that their midpoint would be almost precisely the eccentric anomaly, labeled β in Figure 11. For given β in that figure, the law of cosines yields

$$AP = \sqrt{(1 + e^2 + 2e \cos \beta)}$$

To evaluate the error ensuing from this literal interpretation of the tabulated angles, let us reduce this expression.

$$AP = \sqrt{(1 + 2e \cos \beta + (\cos^2 \beta + \sin^2 \beta)e^2)}$$

$$AP = \sqrt{((1 + e \cos \beta)^2 + e^2 \sin^2 \beta)}$$

$$= (1 + e \cos \beta)\sqrt{\left(1 + \frac{e^2 \sin^2 \beta}{(1 + e \cos \beta)^2}\right)}$$

$$\approx (1 + e \cos \beta) \cdot \left(1 + \frac{e^2 \sin^2 \beta}{2(1 + e \cos \beta)^2}\right)$$

$$AP = 1 + e \cos \beta + \frac{e^2 \sin^2 \beta}{2(1 + e \cos \beta)} + \text{(terms in higher powers of } e) \tag{11}$$

Of course, the equations of center in terms of β also take a different form than (10). In Figure 12, it is clear that the optical equation ϕ is given by

$$\phi = \tan^{-1}(AE/EP)$$

$$\phi = \tan^{-1}\left(\frac{e \sin \beta}{1 + e \cos \beta}\right) \tag{12a}$$

Similarly, the physical equation ψ is

$$\psi = \tan^{-1}(CF/FP)$$

$$\psi = \tan^{-1}\left(\frac{e \sin \beta}{1 - e \cos \beta}\right) \tag{12b}$$

Now we can evaluate the error introduced by using the table as it is labeled, rather than as it is described and constructed. We have an eccentric anomaly; let us call it x to distinguish it from all the Greek letters. To be precise, we

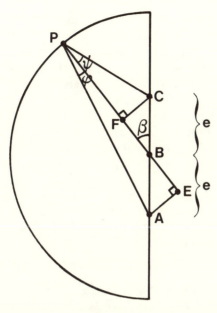

Fig. 12

should obtain the correct distance to the circle by forming the coequated anomaly $x - \phi$, and entering the table where $x - \phi$ is the midpoint between the tabulated angles. This distance, call it R_1, will be that found by substituting $(x - \phi)$ into (8) or (9), but for our purposes it is more convenient to use the equivalent expression (11) involving the eccentric anomaly directly:

$$R_1(x) = 1 + e\cos(x) + \frac{e^2 \sin^2(x)}{2(1 + e\cos(x))} \tag{13}$$

In contrast, suppose that we had taken literally the column headings, and entered where we found the coequated anomaly $(x - \phi)$ in the "coequated anomaly" column. At this point, x would be the midpoint of the tabulated angles. By the construction of the tables, then, we would obtain a distance R_2 determined by the function (9):

$$R_2(x) = 1 + e\cos(x) - \frac{e^2 \sin^2(x)}{2} \tag{14}$$

The error when we took the column headings literally is thus the difference between (13) and (14):

$$R_1(x) - R_2(x) = \frac{e^2 \sin^2(x)}{2(1 + e\cos(x))} + \frac{e^2 \sin^2(x)}{2}$$

For $e = .018$, both denominators are essentially equal to 2 for all x, and so the error at most amounts to about $e^2 = (.018)^2 = .00032$, or 32 parts in a table normed to 100,000. Kepler thus advised the reader that it was safe to enter the table as if it were properly labeled.

We have been a bit cavalier in passing from $(x - \phi)$ to x as arguments of the function. In the discussion leading to (13), the (implicit) tabular equation ϕ was the one computed at coequated anomaly $(x - \phi)$, or eccentric anomaly x. While misusing the table to obtain (14) we used a ϕ which was computed at coequated anomaly x. The difference between the two values of ϕ may therefore be found from the functional expressions (10) and (12a). As may easily be shown,[43] this error in the equations reaches at the octants its maximum of 0;0,33°, so it is safely negligible.

So far we have shown that the curious manner in which Kepler arranged his table did not do too much harm. In order to see why he chose that arrangement we must anticipate some results from his finished planetary theory. First of all, given eccentric anomaly β, the distance to the planet on its elliptical orbit is given by

$$R = 1 + e \cos \beta$$

Comparing with (11), we see that the table, as labeled, will give distances in error by only $e^2 \sin^2 \beta / 2(1 + e \cos \beta)^2$, which has a maximum value of only 16 parts in 100,000. In Chapter 30, Kepler warned that the tabulated distances were as much as 16 parts shorter than the true ones, although they correctly implied an oval orbit.[44]

Likewise, the equations on the ellipse are very close to those in the table as it is labeled. For eccentric anomaly β, the physical equation is just $e \sin \beta$. This is practically identical (within a second)[45] to the tabulated equations (10)—although the latter were calculated as the optical equations on a circular orbit for a different argument, the coequated rather than the eccentric anomaly. Agreement with the optical equations on the ellipse is not quite so good, but the error is never so much as 20 seconds, quite negligible.

These numerical facts make the table of Chapter 30 considerably less bewildering. Kepler must have begun by making the table he promised, giving distances for every degree of coequated anomaly. The distances were those which survive in the final table. There was probably also a column of eccentric anomalies, which survive under the label of "mean anomaly." It was natural

---

[43] In expanding equations (12) we may neglect the inverse tangent function, which for arguments this small is the identity, to five places. We therefore have

$$\phi = e \sin \beta \left(\frac{1}{1 \pm e \cos \beta} \right)$$

$$= e \sin \beta (1 \mp e \cos \beta \pm e^2 \cos^2 \beta \mp \ldots)$$

$$\phi = e \sin \beta \mp e^2 \sin \beta \cos \beta \pm e^3 \sin \beta \cos^2 \beta \mp \ldots$$

The first term is functionally equivalent to (10); the third and later terms never affect the fifth decimal place; and the second term takes its maximum value of 0;0,33° at the octants.

[44] *G. W.*, 3: 229: 36–38.

[45] Again, the inverse sine function in (10) is the identity, to five decimal places, for arguments smaller than $e = .018$.

to tabulate the distances for values of the coequated anomaly, since it was the sun-earth vector that was of interest in his investigation of Mars.

It is obviously not possible that Kepler found the equivalence among the various interpretations analytically, as presented above. There was no need; he would inevitably have come upon them. Upon finding, in 1605, the laws governing the motion of Mars, he immediately extended them to other planets, including the earth. He needed, therefore, a corrected table of the earth's equations and distances from the sun, for the old one had been constructed on the erroneous hypothesis of a circular orbit. As we shall eventually see, the simplest way by far to construct such a table, for Kepler motion on an ellipse, is to enter with the eccentric anomaly. (Indeed, the eccentric anomaly in Kepler motion is not very interesting as an angle. Both the distance and the physical equation, however, have exceedingly simple expressions in terms of the eccentric anomaly.) Constructing such a table, he would have computed the very same values for the mean anomaly which were in his original table as the eccentric anomaly (corresponding respectively to an integral argument of eccentric anomaly, and the same argument of coequated anomaly). He would have calculated distances within 16 parts of 100,000 of his original table, again under a different interpretation of the argument. He evidently saw the resemblances, relabeled the former column of eccentric anomalies, and added a new column of coequated anomalies by simply subtracting the equations he had formerly added.

The final table, then, gave precise results under its original assumption of a circular orbit, provided that one ignored the column headings and did not mind the column of meaningless angles ($\gamma - \phi$). It gave very nearly correct results (within .032%) under the same hypothesis if one accepted the column headings. Finally, it gave an excellent approximation (within .016%) to the truth. The only cost was a very confusing explanation in Chapter 30. And even there, a reader who trusted Kepler's enigmatic advice would not be led astray.

It is clear that the reason this multiple-use table works so well is just that the earth's motion is not very eccentric, and hence can be approximated well with a relatively simple model. Terms in e^3 typically do not even affect a five-place table, and terms in e^2 produce only the negligible errors noted by Kepler. In fact, even the bisection of the earth's eccentricity, important though it was for physical theory and distance calculations, gave no major improvement in terrestrial longitude. Kepler's demonstration of this in Chapter 31 was marred by an error of calculation,[46] but it was nonetheless true. Mars, whose eccentricity was five times as great, was by no means so easily treated.

The New Astronomy

As preparation for the novel ideas he wished to propose regarding the motion of Mars, Kepler devoted the next eight chapters to explicating, as best he could, the fundamental question of what made the planets move in the way

---

[46] Noted by Caspar, *G. W.*, 3: 467.

they moved. Only by investigating the physical causes could he be certain of attaining a hypothesis that was true, rather than merely close to the truth. He attacked the problem by attempting to show how a physical hypothesis, simple and plausible, accounted for the success of the Ptolemaic equant hypothesis.

His physical explanation was that suggested in Chapter 22 of the *Mysterium* (p. 18, above): that the planet moved slower when it was more distant from the sun, in proportion to the distance. In the earlier book he had sketched out an argument that the Ptolemaic hypothesis described motion of just this kind. Here he expanded his reasoning into a geometrical demonstration. To do so, he introduced more rigor into the statement of the hypothesis, phrasing it in terms of the planet's "delay" instead of its velocity. Since Kepler's notion of rigor differs greatly from our own, let us examine the concept of delay, *mora*, a little more carefully.

Velocity, it has been suggested,[47] was not an acceptable tool in Kepler's mathematics, because the velocity of a planet was always changing. Lacking the technical apparatus needed to define instantaneous velocity, Kepler was supposedly reluctant to employ an ever-changing quantity in his calculations. I do not think that this correctly explains Kepler's avoidance of the concept in his mathematics. To be sure, Kepler frequently dealt with problems where a proper theory of infinitesimals would have been enormously helpful. He rarely allowed the deficiency to halt him, however, for he was a tireless and ingenious approximator. (We shall see that he computed sums, and ratios of sums, of ever-changing distances in the course of subsequent calculations from his physics.) If Kepler had wished to specify velocity as a part of his theory, he could likewise have approximated with finite intervals. The likely source of the problem, it seems to me, was not in how to define or use an ever-changing velocity, but in the character of velocity itself as a quotient of unlike quantities, an improper ratio,[48] a ratio that could not be drawn. Velocity was neither a pure number nor a geometrical magnitude. Although Kepler did on occasion solve a numerical problem algebraically,[49] his theoretical analysis was always done with geometry. Kepler could not depict velocities on a geometrical diagram of the orbit.

He could depict arcs and distances directly, of course. Because of the equant circle, on which arcs represented time, he could represent intervals of time. (Later he represented time differently, with areas, but he always did represent it.) His normal practice was to work with the time interval required to traverse a given arc of the orbit, or in his terminology, the planet's delay in that arc. The "delay" was thus almost the inverse of velocity, differing in that he specified what arc was being considered rather than actually dividing by its

---

[47] E. J. Aiton, "Kepler's Second Law of Planetary Motion," *Isis* 60 (1969): 75–90.

[48] The classical theory of ratios is in Book five of Euclid. Concerning the importance of proper ratios to Galileo, see S. Drake in his edition of *Two New Sciences* (Madison: University of Wisconsin Press, 1974), pp. xxii–xxv.

[49] *G. W.*, 3: 95.

length to get inverse velocity.[50] Of course, to make effective use of this rather cumbersome concept he invariably considered delays in arcs that were very small and, in some sense, equal. They were *not* necessarily equal arcs pure and simple. At a more sophisticated stage in his physical astronomy (below, p. 130) he found it convenient to analyze delays in arcs of unequal length. These, however, were equivalent to one another in a sense more fundamental physically.

I shall adopt the language of delays in following Kepler's analysis, occasionally restating things (as did Kepler) in the more intuitive terms of how fast the planet moved at various places. Delays are sometimes inconvenient, but adherence to their use serves the salutary purpose of detaching us from modern notions. It is all too easy to slip into the "correct" way of viewing motion, where velocity persists of its own account, and where one has learned that time belongs in the denominator of ratios if one is to obtain simple expressions for physical laws. In Kepler's world velocity was something active, a struggle between a force and the natural resistance of matter to motion, its inertia. He held fixed the increments of distance traveled rather than those of time because in his own theories time, although still expressible as distances or areas in the geometry of the orbit, could not be divided into equal increments when so expressed.[51] He could divide the distances equally on his diagrams. If we are to understand Kepler's physical astronomy we must learn to think in terms of delay rather than velocity.

Let us now take up the physical interpretation of equant motion in Chapter 32. Kepler's proposition was that in an equant model with bisected eccentricity, "the swiftness at perihelion and the slowness at aphelion are proportioned approximately as the lines drawn from the center of the world to the planet."[52] The proof is sometimes difficult in the text, where it is worked out verbally, so let us run through it. In Figure 13, the solid circle DFEG is the planet's orbit, with B the center of the orbit and A the sun. The dashed circle HKIL is the equant, drawn equal in radius to the eccentric and centered on a point C such that AB = BC = e, the (half-) eccentricity. Clearly the orbit is distant from the equant circle by e at aphelion and perihelion: DH = EI = e. Line FAG, passing through the sun, cuts off the small arcs DF and EG on the orbit at aphelion and perihelion, respectively. These arcs are unequal, but subtend equal (vertical) angles at the sun. To measure the delay in these arcs, that is, the time required to traverse them, draw CF, extending it to meet the equant

---

[50] A more correct inverse for *mora* would be the distance traveled in a specified time interval: not distance per time interval, a quotient, but simply distance, with the time understood. Kepler occasionally used *motus* in this quantitative sense.

[51] This problem remains with us in "Kepler's equation," which expresses time as a function of eccentric anomaly and which is not invertible with elementary functions.

[52] *G. W.*, 3: 233: 36–234: 2. Note that Kepler does not object to speaking, picturesquely, of the planet's *celeritatem in perihelio*; but this phrase is not invested with any precise meaning. The geometrical proof which follows, and which is set off in italics, contains no mention of *celeritas* or anything similar.

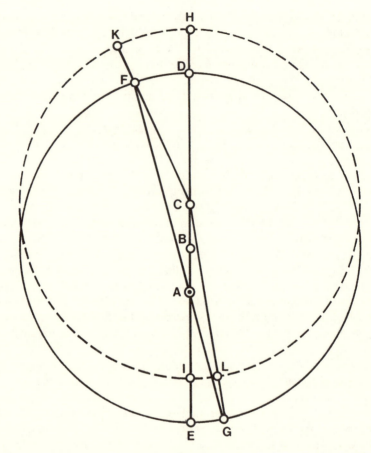

Fig. 13

circle at K, and draw CG, intersecting the equant circle at L. Where the whole circumference of the equant circle represents the planet's periodic time, HK is now the delay in arc DF, and IL the delay in arc EG.

Since we are considering very small arcs, DF and EG on the orbit, and HK and IL on the equant, we may regard them as line segments. Because of the vertical angles at A, the right triangles DFA and EGA are similar. Therefore,

$$\frac{AD}{AE} = \frac{DF}{EG} \tag{15}$$

Also, the right triangles DFC and HKC are similar, as are EGC and ILC. Thus

$$\frac{CH}{CD} = \frac{HK}{DF} \tag{16}$$

$$\frac{CE}{CI} = \frac{EG}{IL} \tag{17}$$

Now, if CH, the radius of the circles, were the geometric mean between AD (radius plus eccentricity) and CD (radius minus eccentricity), then by definition one would have

$$\frac{AD}{CH} = \frac{CH}{CD}$$

Instead CH is the arithmetic mean. Arithmetic means are always greater than geometric means, but where the extremes do not differ by much the two kinds of means are very close. It is very nearly true, then, that[53]

$$\frac{AD}{BD} = \frac{CH}{CD} \qquad (18)$$

where we have inserted BD for CH, which is equal to it.

From (18) and (16) we have

$$\frac{AD}{BD} = \frac{HK}{DF} \qquad (19)$$

In an exactly analogous way, one can treat the radius as a mean between CE and AE, and combine the resulting proportion with (17) to get (approximately)

$$\frac{EG}{IL} = \frac{BE}{AE} \qquad (20)$$

Finally, the radius BE or BD is a mean between AD and AE:

$$\frac{BE}{AE} = \frac{AD}{BE} = \frac{AD}{BD} \qquad (21)$$

By combining (20), (21), and (19), in that order,

$$\frac{EG}{IL} = \frac{HK}{DF} \qquad (22)$$

Thus far, Kepler had considered the unequal arcs DF and EG. But the delays in small arcs at the apsides are proportional to the lengths of these arcs, according to (16) and (17). Each of the ratios in (22) is therefore independent of the length of the (very small) arc of the orbit examined. If these lengths DF and EG were equal to one another, (22) would still hold in the now-equal arcs; and it would state that the arc length is a mean proportional between the two delays. Let us denote the adjusted lengths by DF′ and EG′, where DF′ = EG′ = s, say, and the corresponding delays by HK′ and IL′. We still have

$$\frac{s}{IL'} = \frac{HK'}{s} \qquad (22a)$$

So

---

[53] There is a textual error in the statement of (18). In *G. W.*, 3: 234: 27, read "*Sed γδ est ad γυ...*"

$$\frac{HK'}{IL'} = \frac{HK'}{s} \frac{s}{IL'} = \left(\frac{HK'}{s}\right)^2 = \left(\frac{HK'}{DF'}\right)^2$$

Equation (16), depending only upon similar triangles, still holds for the adjusted arcs HK′ and DF′, so

$$\frac{HK'}{IL'} = \left(\frac{CH}{CD}\right)^2$$

$$= \frac{CH}{CD} \cdot \frac{AD}{BD} \qquad\qquad \text{by (18)}$$

But CH = BD, the radius, so those terms cancel out of this last product; and we may substitute AE for CD, since each equals the radius minus the eccentricity.

$$\frac{HK'}{IL'} = \frac{AD}{AE} \qquad\qquad (23)$$

For equal arcs at aphelion and perihelion, the delays are directly proportional to the distances from the sun. This was Kepler's conclusion: the Ptolemaic equant model implied that delays (at the apsides) were proportional to distance from the sun.

We should remark that Kepler's approximations, in interchanging arithmetic and geometric means, could have been avoided by more elaborate constructions, or by the algebraic techniques he avoided. The "distance law" holds exactly—at the apsides—for equant motion with bisected eccentricity, and, incidentally, for Kepler motion on an ellipse. Kepler himself stated only that it was true *quam proxime*, and probably did not know, when writing the *Astronomia nova*, of its exact validity.

Outside the apsides the theorem is not exact. Kepler remarked this fact in closing Chapter 32, claiming that it was of little consequence. At other places in the orbit, he argued, the extreme terms of his proportions differed by less than the double eccentricity. This meant, of course, that the approximation of arithmetic means for geometric means introduced less error (note that he really did think the theorem inexact). In point of fact, as we have remarked, these approximations were of no consequence. Kepler's proof breaks down outside of the apsides because the radii from the sun and the equant center no longer intersect the circles at right angles, so that the three pairs of triangles used in obtaining (15)–(18) are no longer similar.

In later chapters Kepler did use the "distance law" for all parts of the orbit. This was not, surely, because he forgot that he had only proved it at the apsides. He used it because he thought it was physically true. He had shown that the equant model, a geometrical expedient, was approximately true—that is, almost equivalent to the physically true theory—both at the apsides and elsewhere. It followed from this demonstration, of course, that the epicyclic model Copernicus had used instead of the equant was also very nearly equivalent to Kepler's physical theory. His purpose in Chapter 32 was to explain how the equant could have worked as well as it did,[54] just as his purpose in Chapter 5 (pp. 35–38, above) had been to show how models based on the mean sun could have been approximately true. In each case the near-agreement of his physical theory with older geometrical theories was merely a part (although an essential part) of the more general argument that the theories following from physics were really true. The equant model, like the models Kepler had so casually spun out when explaining how a false theory could produce apparent truth, seemed accurate; but to ascertain the true motion of the planets one had to understand the causes of their motion.

Kepler had demonstrated at last, although not with any great precision, what he had been asserting all along: that all adequate theories of the first anomaly tacitly implied a remarkable correspondence. "When the planet is farther from that point which is taken for the center of the universe, it is moved more weakly around that point."[55] It was time he followed up this clue.

Distance between the bodies, he argued, clearly took causal priority over their circulation, and if the changes in motion originated from changes in the distance between the two bodies, a cause of the motion must be present in one or the other of them. Nor was it difficult to choose one for closer consideration. The system's central body, in any of the world hypotheses, participated in all of the distance-motion relations. Supposing that this central body was the source of whatever the motive virtue might be, then it was evident that the action of the virtue was weaker at greater distances. This seemed to Kepler an entirely plausible explanation, since there were well-known instances, in the lever and in scales, of the effect of a motive force being weakened in proportion to the distance at which it had to operate. The relation between distance and velocity was surely the explanation for the planet's changing speed.

This was a physical explanation. An "animal" force, that is one arising from some kind of soul in the planet, would have worn out if it had to vary its intensity so, Kepler argued. Furthermore, planets were round, and not equipped with wings or legs by which a soul might move them from one place to another. It was no soul that moved Mars around the sun, but a physical force. Variations in velocity were not to be ascribed to inexplicable changes in the activity of a soul, for they patently followed from the easily understood effect of increased distance.

These simple physical arguments had cosmological implications that told heavily in favor of a (modified) Copernican system, where the central body causing all the motion could be identified with the sun itself. The Ptolemaic models all lacked a body at the epicyclic center, and hence could not account for the changes Kepler had found in the planet's speed on the epicycle. Tycho's

---

[54] Aiton, "Kepler's Second Law," p. 78, has appreciated the fact that in this chapter it was the equant, and not the distance law, which Kepler thought to be an approximation; also "Infinitesimals and the Area Law," in Internationales Kepler-Symposium, Weil-der-Stadt 1971, p. 293.

[55] G. W., 3: 236: 12–13.

system was somewhat better, since most of the planets circulated about the sun in a reasonable way. There was no prospect, however, of giving a credible account of the entire system's supposed motion around the earth. For all its geometrical equivalence to the Copernican system, Tycho's ill-proportioned hypothesis was as unlikely in Kepler's physical world as in Aristotle's, or in ours.

(Admittedly, a similar problem arose for Kepler on a smaller scale, for the moon evidently circled the earth, which therefore had to be the seat of some secondary impelling force. The moon's motion, however, was a much more modest phenomenon. Furthermore, a closer examination revealed that it too was tied to the solar body, the source of all local motion in the heavens. We shall see how Kepler dealt with these problems below on pp. 175 ff., in our discussion of the lunar theory in his *Epitome Astronomiae Copernicanae*.)

The motive virtue, then, was surely resident in the sun. It seemed to share a suggestive characteristic with light: it emanated through space from the sun without suffering any loss. In Chapter 32 Kepler had shown how the delay of a planet in some small arc of its path increased in proportion to distance. But the circumference of an entire revolution about the sun increased in this same proportion: so that the capacity of the solar virtue to produce motion was the same, in total, at any distance.[56] None of the virtue was lost in transit. The same held true of light, he remarked, as he had demonstrated in the first chapter of his book on the *Optical Part of Astronomy*.

Kepler's elucidation of the sun's virtue, or capacity, to move the planets drew extensively upon this analogy with light. Neither light nor the virtue was dissipated between sun and planets; therefore "just as light ... is an immaterial image of that fire which is in the body of the sun, so this virtue, grasping and carrying the bodies of the planets, must be an immaterial image of that virtue which resides in the sun itself...."[57] Again citing the first chapter of his *Astronomiae Pars Optica*, Kepler further proposed that the virtue, or image, was no geometrical body, but was like "a kind of surface, just like light...."[58] It was received by, and it acted upon, the whole volume of the planets, but in itself it was best considered as a surface, like light.

This passage has been almost universally misinterpreted as meaning that

---

[56] The circumferences Kepler spoke of in this argument (*G. W.*, 3: 239: 9–16) were not of any real orbits, but rather of imaginary paths centered on the sun, which a planet would follow if no other force affected its distance from the sun. Notice that for Kepler no centripetal force was needed to bend the orbit into a circle, since he did not think that the planet's inertia kept it moving on the tangent.

[57] *G. W.*, 3: 240: 19–22. "Image" is our rendering of Kepler's *species*, which has for the most part been left untranslated in other accounts. As Kepler used it the word seems to mean the appearance or visible manifestation of the sun; we shall present evidence for such an interpretation. Meanwhile we will use the English word "image" for concreteness. There should be no confusion with *imago*, an optical image, a word which Kepler used in his optics but not in his astronomy.

[58] *G. W.*, 3: 240: 28–29.

the solar image, or *species*, was confined to the plane of the ecliptic.[59] That interpretation is superficially all too plausible, for a couple of reasons. First, Kepler stated explicitly that the virtue is like a surface. Second, confining the virtue to the ecliptic plane accounts for its attenuation in only the simple inverse proportion to distance. Something emanating spherically must attenuate as the inverse square of distance, it seems, for the areas of the spheres over which it is spread increase as the square of distance. On the other hand, circular circumferences increase only in direct proportion to the radius, so that we can easily understand how something expanding outward in a plane could fall off in strength by the simple-inverse proportion. This latter reason, I think, is the chief culprit. Since Newton, Kepler's failure to use an inverse-square law for his solar force has been taken as the chief problem in understanding his physics.

It is a false problem, based upon a force comparison with gravitation, and it has blinded readers to what Kepler actually wrote. He did not write that the virtue was like a plane, but like a surface (*superficies*), and surfaces need not be planes. Moreover, he said that it resembled a surface "just like light" (quoted above). To interpret this we must turn to the first chapter of Kepler's *Astronomiae Pars Optica*, a chapter he cited twice in these passages. There we learn that the motion of light is in straight lines, but that what is moving, the light itself, must instead be considered as a surface (*superficies*) expanding spherically and with infinite speed from the origin of the light.[60] Kepler carefully distinguished between the motion (*motus*) of light in straight lines, and the surface formed by that which moved (*mobile*). Expanding through space, the surface coincided with what we would call a wavefront;[61] at its destination it coincided with the surface illuminated.

We need not digress into Kepler's theory of light. It is clear enough that he likened the solar virtue to a *spherical* surface because of the way it expanded through space, like light, suffering no dissipation. He did distinguish the action of light in illuminating a surface from the action of the motive virtue throughout the "corpulence" of the bodies it moved. The action of light on a surface was instantaneous, while the motive virtue acted in time because of the resistance of the body on which it needed to act. Perhaps, he suggested, this action was more like the bleaching of colors by light, a process within the matter which took time to occur.

This interpretation of the "surface" formed by the solar image is surely the correct one. It does not by any means resolve all the obscurities in the action of the solar image. In these same pages Kepler phrased the following contrast:

[59] For example, J. L. E. Dreyer, *A History of Astronomy from Thales to Kepler* (New York: Dover, 1953), pp. 387–388; M. Caspar, *Kepler 1571–1630* (New York: Collier, 1962), p. 143; A. Koyré, *The Astronomical Revolution* (Ithaca: Cornell University Press, 1973), p. 409; Aiton, "Kepler's Second Law," p. 78, and "Infinitesimals," p. 294.

[60] *G. W.*, 2: 21–22.

[61] Kepler did not think of light as waves.

"light flows out in straight lines spherically, the moving virtue in straight lines also but circularly; that is, it presses in only one direction of the heaven, from west to east, not backwards nor toward the poles, etc."[62] This difficulty in his analogy he postponed to Chapter 36 (see pp. 72 ff., below), along with the related one that light was attenuated as the inverse square of distance.

Kepler put off discussing the solar image because he needed to develop some probable conclusions about the sun, its source. His first point was that the image moved in the direction toward which it impelled the planets. Yet it was immaterial (since none of it was dissipated), and hence it propagated instantaneously. Its motion must be tied to a motion of its source, so that each part of the image retained the motion of the part of the sun from which it originated. The sun, therefore, rotated along with the planets.[63] As the sun rotated in place, its image swept around, as shown by the example of an orator rotating his head in the middle of a crowd of listeners. Where he faces, "they see his eyes," while elsewhere "they lack the aspect of his eyes." As his head turns the aspect of his eyes sweeps through the crowd.[64]

The instantaneous propagation of the image did not by itself move the massive bodies of the planets. The rotation of the image, on the other hand, was of finite velocity, as it was tied to the rotation of its material source. The virtue moved the planets because of this rotation, and hence because its source moved. The sun's rotation, in this theory, was ultimately the cause of the movement of all the planets around it. Midway between the poles of the sun's rotation lay the great circle of the ecliptic, whose natural origin had previously been unknown.[65] Near the plane of the ecliptic moved all the planets, with the more distant moving more slowly. They obviously were not all moving at the speed of the image; so evidently they resisted its impetus, having material bodies which were "prone to rest or to the privation of motion."[66] The sun, then, rotated faster than any of them. Kepler speculated as to its period. The analogy sun: Mercury: :earth: moon led to a period of three days; but perhaps, he mused, the earth's twenty-four hour period was more appropriate. In any

---

[62] *G. W.*, 3: 240: 8–10.

[63] As is well known, Kepler reached this conclusion several years before Galileo observed the motion of sunspots. I suppose that, in a very general sense, his reasoning was sound: the entire solar system rotated in the same direction, so it was physically likely that the sun did likewise, whatever the forces connecting sun and planets.

[64] *G. W.*, 3: 243: 9–15. In view of some statements in Chapter 36, it may be that the visual terminology in this metaphor is not accidental. The word "aspect," used both here and in that chapter, is interesting because of its prominence in Kepler's astrology, and because it did not commit him to saying whether the metaphor was about the sun seeing the planets, or the planets seeing the sun.

[65] Delambre (1: 439) objected to this statement, pointing out that Kepler had no justification for identifying the solar equator with the orbital plane of the earth. Kepler obviously knew this; indeed he was the first astronomer to consistently treat the earth as a planet like the others. Delambre's oversight indicates only that he failed to study Part V of the Astronomia nova. In fairness, I know of no one else who has studied those final chapters.

[66] *G. W.*, 3: 244: 19–21.

case, he was confident that the sun rotated on its axis, and in a time somewhat less than Mercury's ninety-day period.

Speculating on what the solar body could be, that it rotated and exerted a force on the distant planets, Kepler switched to a further analogy, with the magnet. The attractive force of a magnet was thrown out spherically (*orbiculariter*), he remarked,[67] creating a sphere of attraction within which iron was attracted, and attracted more strongly when closer to the magnet. "Clearly in the same way," he continued, "the virtue moving the planets is propagated from the sun *into a sphere*, and for the more remote parts of that sphere it is weaker." [68] Indeed, the sun did not attract the planets to it, as a magnet attracted iron. However, Gilbert had carefully distinguished a second faculty in the magnet, that of directing iron parallel to the magnet, as seen in a compass. If one supposed that the sun contained fibers parallel to its equator, and endowed with a directive power like that of the magnet, then their image might somehow carry the planets as the solar fibers were themselves carried by the sun's rotating body.

Best of all, Gilbert had demonstrated quite recently that the earth itself was actually a giant magnet. If it had circular fibers also, around its axis, one could explain the lunar motion by the same theory as that of the planets. Granted that the moon stayed near the ecliptic, while the earth's fibers would be parallel to its equator, all other things seemed agreeable to this notion. The earth was as intimately tied to the lunar motions as the sun to the planetary motions, and both kinds of orbit had similar, periodic inequalities. Surely the earth was as responsible for the one as the sun for the other. And if the earth moved the moon and was a magnet, should we not think the sun, which moved the planets, to be also a magnet?

At this point (Chapter 35), Kepler briefly raised the question whether planetary occultations would block the solar image, and if so whether the very slow motion of the apsides and nodes could be attributed to such occasional lapses. His excitement at the power latent in physical explanation outstripped his theory here, and he was not able to answer either question with any assurance. The analogy with light was perhaps unreliable in this instance, he decided, since it was opacity rather than body as such which blocked light. Nor was the analogy with magnets helpful, for their force passed through most substances (silver, copper, gold, glass, bone, stone) but was blocked by the interposition of a magnet tablet (*magnetica tabella*), and likewise by an iron tablet. These evidently were able to "drink up" the magnetic virtue. The case of the planets could fall either way, on this analogy, depending upon whether they were sufficiently like the sun to be able to drink up its virtue. The question was dropped with a promise to resume it in Chapter 57 (pp. 110 ff., below).

Kepler turned again to the problem of justifying the simple-inverse proportionality of the motive virtue with distance. Light, as he well knew, was

---

[67] Probably following Gilbert, *De Magnete*, Book 2, Chap. 7.

[68] *G. W.*, 3: 246: 1–3 (my emphasis). Note that the virtue is explicitly not confined to the ecliptic.

attenuated as the square of distance from its source.[69] Evidently collecting all the objections he could think of, Kepler proposed in hypothetical and rather obscure arguments that both light and the motive virtue must weaken as at least the square, and perhaps even the cube, of distance from their source. He then refuted all of these arguments, either as based on inadmissible ratios of quantities equal to zero, or as relating merely to the apparent size of the source.

The proper way to answer this question, Kepler insisted, was to avoid all complications relating to the size of the solar disk, or to its appearance. The total amount of light, or of the virtue, was the same on a spherical surface of any radius. Either was attenuated only as these spheres expanded, he stressed, and hence in inverse proportion to distance.[70]

This explanation raises more questions than it answers. How did Kepler conclude that the expansion of the sphere weakened the virtue only in simple inverse proportion to distance? He knew that the surface of a sphere was proportional to the square of its radius, and had stated as much on the previous page. For that matter, how could an argument, phrased both in terms of light and of the motive virtue, distinguish between their different modes of attenuation? To the latter question we can only demur that the statement of simple inverse proportionality evidently applied only to the motive virtue. The former question, which has been with us all along, we shall attempt to unravel with the aid of some curious remarks Kepler made in closing the chapter.

It might be thought, he cautioned, that the total amount of virtue would not be conserved in successively larger spheres. Light, which was conserved, emanated spherically; but the virtue was needed only in the vicinity of the ecliptic, where the planets moved. Indeed this has been the opinion of most who have written upon Kepler's physics; not because they doubted the conservation of solar virtue, but because that is the easiest way to visualize a simple-inverse attenuation. Even a casual reading of the final page of Chapter 36 renders this interpretation untenable. Because of the persistence of this misreading, I translate at length. The "filaments" (*filamenta*) Kepler speaks of here are imagined as fibrous structures encircling the sun like parallels of latitude.

> It is replied, that the case of light and of the motive virtue is entirely the same, and the deception is in the reasoning. For just as in light the rays do not flow out from the points and circles of the body of the sun to the corresponding points and circles of the sphere ... but the rays emanate from an entire hemisphere of the shining body to every point of the imagined spherical surface...: so also in the matter of the virtue the same thing holds. For even though the magnetic filaments of the solar body are arranged following the length of the zodiac; and even though only one great circle of the body of the sun lies under the zodiac or

---

[69] For example, *Astronomia Pars Optica*, 1, prop. 9, in *G. W.*, 2: 22.
[70] *G. W.*, 3: 250: 19–27.

a   b   c

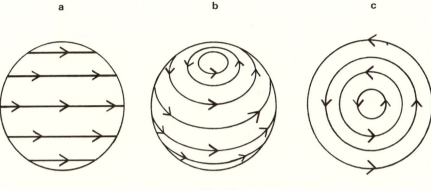

Fig. 14

ecliptic, and approximately the orbit of the planet; lastly, even though the other, lesser circles (decreased at last to the smallness of a point at the poles) are arranged under their corresponding circles in the sphere of the planet: yet the rays flow out from all the filaments of the solar body (originating from one hemisphere of the body) and converge not only at all the points of the path of some planet, but at the very poles over the poles of the body of the sun. And the body of the planet is carried in proportion to the density of this whole image, composed out of all the filaments.

But it does not follow from this, that just as the sun shines equally in all directions, the planet also, as you might fear, would move indiscriminately in all directions. For the magnetic filaments of the sun do not move it when they are considered in isolation, but only insofar as the sun, turning around rapidly in its place, carries around the filaments also, and with them the moving image spreading from them. The planet, therefore, will not go backwards, since the sun always rolls around in the forward direction [*in directum*]. The planet will not go toward the poles (even though some of the image of the sun's body is in those points also): since the filaments of the solar body are not directed toward the poles, nor does the sun rotate in that direction, but in that toward which its filaments draw it.[71]

This passage establishes conclusively that the solar virtue was *not* confined to the ecliptic. It propagated in all directions; only its motion was circular.

Kepler went on to explain how it came about that the planets' motion remained near the ecliptic. To a planet near the zodiac, the solar body looked like Figure 14a. It was encircled by quasi-magnetic filaments or fibers (*fibrae*).

---

[71] *G. W.*, 3: 251: 3–30. On 251: 22 read *Planeta*, with the 1609 text, rather than *Planetae* as printed in *G. W.*

Those parts of the magnetic fibers that were presented to it—that is, the parts that were visible—were semicircular in form, but appeared straight, moving in concert from left to right across the face of the sun. Their image impelled the planet in the same direction. If, on the other hand, the planet were at some appreciable latitude, the fibers in the hemisphere toward it would be arranged as in Figure 14b. Most of the visible fibers move from left to right, as before. However, parts of some of them are beyond the pole, and are moving from right to left. Of these fibers, Kepler asserted that "the image will be composed of filaments tending in contrary directions." Such an image is "less suitable ... for giving motion to the planets."[72]

It seems clear enough what Kepler meant here. He had always emphasized that it was not the mere presence of the *species*, but its motion which impelled the planets. This motion derived entirely from the rotation of the solar body. So far as concerned a planet in the plane of the solar equator, the sun's motion was quite unidirectional. That is, the appearance or visible manifestation of the motion was altogether in the direction designated "east." From a higher latitude the motion appeared more complex. Extrapolating to the view from above the pole (Figure 14c), the appearance of the sun's motion would have tended equally in all directions. The planet, which was carried by "the whole image, composed out of all the filaments," would thus have been impelled in all directions equally, and would not have moved.

Kepler produced this extraordinary bit of exposition to explain why all the planets were nearly in the same plane. It is not clear that his argument accounted for that phenomenon, at least not in physical terms. What the argument does reveal, of interest to us, is that Kepler used the word *species* in the literal sense of something seen, an appearance or image, and not in some obscure metaphysical or mystical sense. He did not invent an explanation of how the planet was able to perceive the image of the rotating magnetic filaments; that was merely a supposition he thought likely. In light of this last argument, though, there can be little doubt that his choice of a term relating to vision, *species*, was conscious.

We are now in a position to evaluate Kepler's justification of the simple inverse attenuation of the moving image. This question is important, we reiterate, not because of the anachronistic contrast between Kepler's solar force and Newton's gravitation, but because of Kepler's own analogy between light and the motive virtue.

(Before proceeding, we should emphasize that Kepler's justification of the simple inverse proportion was not at all the same as the real reason he accepted it. He believed that proportionality for more basic reasons. The planets varied their velocity in inverse proportion to distance from the sun: hence that which moved them followed the same proportion. Assuming as he did that bodies remained at rest unless moved, Kepler was given the simple

---

[72] *G. W.*, 3: 251–252.

inverse relationship as a manifest empirical fact. In the following decades, as the self-evidence of his assumption eroded, the empirical aspect of the distance law was lost. Kepler's own justification, deprived of its original support in the "facts" of astronomy, was long treated as a historical curiosity, its own structure lost in the radiance of Newtonian physics. More recently, the facile reading of "surface" as "plane" has blocked whatever insight might have been gained from his original articulation of the theory.)

The crucial elements in his account were the two analogies, with magnetic influence and with light. The magnet was an important analogy, obviously, because it acted upon other bodies physically removed from it, demonstrating the possibility of physical influence from a distance, through empty space. Equally important, and essential to the analogy, was the magnet's faculty of "direction." As that great magnet the earth directed a compass needle, so the circular fibers girding the sun directed the motion of the planets. The solar force was not the same as the magnetic, for it did not attract the planets; instead it moved them by the motion of its filaments.

For the analogy with light, I think we must again take up Kepler's statement of Chapter 33: "light flows out in straight lines spherically, the moving virtue in straight lines also but circularly...." Comparing with the account of light in the *Astronomiae Pars Optica*, we see that the adverb *orbiculariter*, spherically, describes the form assumed by the light itself. Likewise, I think, we are to take *circulariter* as implying that the motive image retained the form of that whose image it was, the circular fibers or filaments parallel to the equator of the sun. The image was not material, and perhaps one should think of it as a kind of energy, like light. Moreover, it moved, with the motion of the solar filaments, and it was only by this motion that it carried the planets. Now, Kepler was not at all abashed in explaining that the image was present at all points of a sphere centered on the sun, and that in larger spheres it was attenuated only by the expansion of the sphere, namely in inverse proportion to the distance. To reconcile these assertions, we must recognize the image as taking the form of a rotating circle. The *motion* of the image loses its efficacy in proportion as the circumference of the circle stretches, and thus in simple proportion to distance from the sun.

In view of Kepler's discussion of the image at high latitudes, we must not associate any of the circular images with specific circular filaments. Each part of the image was "composed out of all the filaments," and its motion was likewise a composite of all the motions which were visible, that is, all of the motions of the filaments in the visible hemisphere of the sun. At any point in the solar system, the efficacy of the image in moving the planets came from the sum of all the motion visible from that point. The composite motion was directed along a circle parallel to the sun's equator (the resultant of the images of all the filaments), and it was therefore weakened— at any latitude—only as this circle expanded, in the simple proportion of distance.

Libration: Motion in Altitude

The moving image, if it alone had acted upon the planets, would have carried them around circles centered on the sun and in the plane of the solar equator,[73] with constant velocity. Planets did not move so simply. Each planet was sometimes closer, sometimes farther from the sun. Consequently, the speed with which it was carried around varied, in inverse proportion to distance. Moreover, the planets moved in different planes, apparently escaping somehow from the sun's equatorial plane. Beyond the common movement around the sun, then, each planet showed two periodic oscillations in directions perpendicular to that of the moving solar image. These could not arise from the sun's moving image, which was simple and unvarying, although its constant strength was weaker at greater distances. Kepler concluded that each planet required its own innate force, a *vis insita* to move it or steer it—he was not certain which—through its wandering course.

Each planet had a different eccentricity, and each moved in a different plane. These motions did not interfere with one another. To understand such motion, Kepler needed to study the action of the *vis insita*. He began in Chapter 39 by scrutinizing the changes in the planet's distance from the sun, or as he sometimes said, its altitude above the sun.

Kepler imagined the planet as located in a moving *radius virtuosus* which

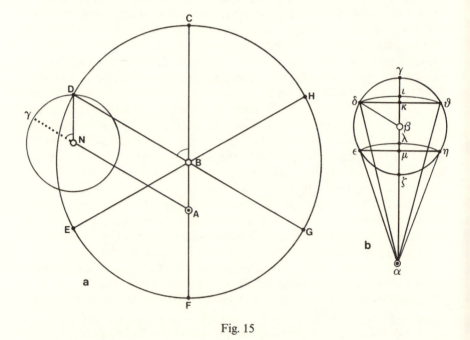

Fig. 15

---

[73] Or in a plane parallel to the equator, if the initial position had been outside the equatorial plane.

joined it to the sun. Motion toward or away from the sun he thought of as occurring along this radius, which served as a *locus*, or frame of reference, for the oscillation. Each of the inequalities, in distance and in latitude, had a unique direction and quantity for each planet. Such diversity, Kepler argued, could not arise from the simple and constant action of the solar virtue. The planetary inequalities must have causes specific to each planet. It was as if a planet had a pair of rudders by which it steered in and out, up and down in the steady stream of the sun's moving image. Kepler set out in Chapter 39 to analyze the shape of an orbit in physically meaningful coordinates, so as to determine how the virtues proper to the planets acted.

In this analysis he used two fundamental diagrams, those of Figures 15a and b. In 15a, the planet's orbit, assumed to be circular, is represented with its center at B and the sun at A. The equivalent epicyclic representation for a select position D appears also on the same diagram. The epicyclic center N moved on a circle (not shown) concentric to the sun, and the planet moved around the center N, in the opposite direction, so that ND was always parallel and equal to the eccentricity AB of the eccentric. If AN is extended to γ, the angle γND of anomaly on the epicycle is equal to the angle CBD of eccentric anomaly. All of this follows immediately from consideration of parallelogram ABDN. This was the standard representation of the shape of a planet's orbit,[74] but Kepler wanted particularly to analyze the distances from the sun A to various points on the path. In order to compare these distances more conveniently, he transferred the sun and epicycle to diagram 15b. Here α is the sun and β the epicyclic center. To preserve the equivalence, αβ is equal to the radius of the eccentric in Figure 15a, and the epicyclic radius is the same as the eccentricity. We may now transfer any distance to Figure 15b by marking off an angle of epicyclic anomaly equal to the eccentric anomaly in the orbit. For example, the eccentric anomaly at point D is measured by angle CBD, so we construct angle γβδ equal to CBD. By construction, then, the triangles AND and αβδ are the same, so that αδ equals the planet's distance AD from the sun. (It is obviously immaterial to which side of the line αβγ we make this construction.) For direct comparison of distances, one can simply transfer them all to the central line by drawing an arc centered on α, for example, the arc δι.

The problem was easily visualized on these diagrams. As the planet moved from C to D, E, and F, its epicyclic representation moved from γ to δ, ε, and ζ. The corresponding distances were αγ, αδ = αι, αε = αλ, and αζ. The sim-

---

[74] Notice that the equant is not needed in the present discussion. Kepler had attributed the "physical inequality" to the weakening of the solar virtue at great distances. Therefore he could (and consistently did) ignore it entirely in his analysis of the physics underlying the shape of the orbit, for once the shape was correctly determined, the physical inequality took care of itself. Isolating the optical part of the inequality was a great convenience for Kepler. As a consequence, of course, time or delay was not available for use in his analysis of the planet's particular motions. He treated these motions, as he had treated the delays, as functions of the distance traveled by the planet.

plest way Kepler could imagine for the distances to vary like this was for the planet, or some virtue proper to it, to explicitly trace the circle or circles by which its path was constructed. This possibility he had rejected, for the reasons given in Chapter 2 of the *Astronomia nova*. Many absurdities were involved in supposing that a planet could move, and move non-uniformly, in a circle about an empty point, which was itself moving, non-uniformly, in a circle about the sun; or that it could move, non-uniformly, about the vacant center of the eccentric, with no guide except the apparent magnitude of the solar disk. Such complicated hypotheses, although designed to yield a perfectly simple eccentric circular path, were not physically credible.

It seemed to Kepler, therefore, that the planetary mover surely concerned itself only with the proper regulation of its distance from the sun, disregarding entirely any eccentrics or epicycles astronomers might use to construct an equivalent motion. He could construct the amounts of approach and withdrawal, as we have seen, but he could find no obvious pattern in them. In Figure 15b it is clear that the distances by which the planet approached the sun, while it traversed the equal arcs γδ, δε, and εζ of its orbit, were respectively γι, ιλ, and λζ. These increments were small near the apsides, and larger in the intermediate distances. Moreover, an increment near perihelion, as λζ, was larger than the corresponding increment γι near aphelion. Motion of such irregularity was controlled neither on the basis of the planet's own delays, for the delays corresponding to equal arcs decreased throughout the semicircle; nor on the basis of the planet's angular motion about the sun, for the angles motion increased (as angle DAE, Figure 15a, is greater than CAD, and EAF greater than DAE). The peculiar pattern of the planet's motion toward the sun, as it moved on its eccentric circle, was regular and cyclic, and demanded physical explanation. There was no analogue, however, no physical quantity that revealed how the pattern was maintained.

Notice the remarkable thing that Kepler was doing here. He was analyzing motion on an eccentric circle, a model that had been in general use for nearly two millenia, apparently the simplest possible model with any empirical accuracy. He took apart this beautifully simple model and showed that as a physical process (and in the absence of solid spheres) it was really quite complicated, so complicated as to raise doubt about whether it could be real. He had performed so radical a reassessment by interpreting astronomy, for the first time, as a physical science. From this approach he gained very little in the use of existing physical theory; instead, he found novel and effective criteria for evaluating theories. No longer did it suffice that a theory was mathematically plausible. Mathematical elegance appealed to the astronomer, but real bodies were moved by physical forces acting on other real bodies. The convenience of the astronomer yielded to the constraint of objectivity, as Kepler resolutely concentrated on the physical world.[75] In Chapter 2 (above, pp. 27 ff.),

---

[75] Kepler also believed that the natural world, because it had been created, was organized according to principles more fundamental than mere physics. His astronomical work, however, was always directed at finding the order that was objectively within nature.

he had argued that the intrinsic simplicity of eccentrics and epicycles was irrelevant, for Tycho had disproved their solidity, and insubstantial circles could not constrain the motion of a body. Here in Chapter 39, he analyzed them into components with physical significance, in effect the components of a heliocentric spherical-coordinate system. Showing that the component of motion toward the sun was unexpectedly complex, he extended his earlier critique by raising the question whether "simple" eccentric motion was possible even as a secondary phenomenon, due to other causes.

Kepler was unable to see how an eccentric circular orbit was physically possible. Nevertheless, he put aside that problem[76] for another one, one more fundamental in his physics. In some unknown way the planet determined how far it needed to approach the sun, or withdraw from it. (Kepler called this motion toward and from the sun "libration," evidently with reference to the back-and-forth motion of a balance, *libra*.) The need remained for a standard of comparison. Supposing the most generous possibility, that a mind associated with the planet could remember or calculate the quantities of libration needed to keep the orbit circular, even so the mind would need to know its relative distance from the sun. Without this, regulation of the libratory motion would be impossible. As so often, appreciation of this argument depends upon our thinking about motion the way Kepler did: the planet's motion was forced motion, against the resistance of inertia. If the planetary mind, or whatever exercised control, initiated no action, no change in distance would occur. To achieve the correct change in distance without comparing the initial and final distances would have been to measure without a rule. (A regulatory mechanism that was non-mental would likewise require information concerning the distances. Kepler was considering an extreme case, as he usually was when he spoke of planetary minds.) Some measure of distance, indeed, might prove necessary even for a theory less involved than the eccentric circle had turned out to be.[77]

Changes in the planet's distance from the sun did reveal themselves in one physically prominent way. The apparent size of the solar disk varied with any change in distance, and provided at least a possible link between the motive virtue proper to the planet and the essential information about distance. Kepler was not, he hastened to add, postulating that planets can see as people see. The fact was that we could not know how many ways of perceiving there might be in nature. The mode was unknown by which astrological aspects influenced people's lives, for example, but that did not deny their evident influence.[78]

For the time being, then, Kepler would assume that the planet could

---

[76] *In penuria melioris sententiae, G. W.,* 3: 260: 5. By the time he published the book, he was able to add a promise here of observational evidence that the path was in fact not an eccentric circle.

[77] *G. W.,* 3: 262: 31–36.

[78] Kepler also adduced his successful theory of latitude as evidence that the planets, or their movers, took account of the sun's position. A sentence probably added later discounted the relevance of perception to motion in latitude, on the basis of Kepler's physical theory of latitude.

approach or withdraw from the sun, in some way as yet unspecified, so that the apparent solar diameter was about in the inverse proportion of the distance to the correct point on the eccentric. Before he could say anything more precise, he had to specify mathematically the physical theory obtained thus far.

In broad outline, the theory was this: the optical part of the equation of center (that is, the apparent change in motion due to the orbit's being eccentric to the sun) arose from some virtue proper to each planet, which alternately moved the planet toward and from the sun, in such a manner as to define the shape of the orbit. The physical part of the equation (the actual variation in the planet's rate of motion around the sun) was a secondary effect, realized only as the planet's libration carried it to distances where the moving solar image was stronger or weaker. Calculation of the optical equation was just trigonometry, given the shape of the orbit. The physical equation posed new problems, for the distance law was stated only in terms of small increments of motion. The delay, or time, required to traverse a small arc was proportional to the planet's distance from the sun. This distance was changing all the time, and Kepler lacked the formalism of the integral calculus to tell him how he might accumulate the delays.

He first tried to accumulate them by brute computation. Dividing the circumference of the circular orbit into 360 equal arcs, he calculated the distance from the sun to each of the arcs. Since this distance changed very little within a one-degree arc, Kepler supposed each distance to be proportional to the delay in that arc. The sum of the distances, then, had that same proportion to the periodic time, which was known and conventionally designated by 360° of mean anomaly. These calculations solved the problem. Kepler computed the sum of the distances to each one-degree arc in the orbit, and he computed all the partial sums of the distances to each arc, from aphelion to any given point in the orbit. The mean anomaly at any point in the orbit was to 360° as the sum of distances to the same point was to the overall sum of distances. The method was very tedious: finding the mean anomaly at one place involved nearly as much labor as constructing a complete table. Kepler looked around for a better way.

He recalled the technique of Archimedes, whereby a circle was divided heuristically into an infinity of triangles to determine its area. Thus in Figure 16 one could divide the circular area into sectors from the center B, sectors such as CBG, GBH, etc. If the divisions were made very fine, each sector was very nearly an isoceles triangle; and the bases of the triangles together made up the circumference of the circle. If one consolidated some of these triangles, the combined area was obviously proportional to the arc formed by the connected bases, thus to the change in eccentric anomaly. Kepler also suggested a different way of looking at the area. The radii from the center (infinite in number) all lie within one or another of the sectors. It is more or less obvious—Kepler did not try for rigor here—that a bigger sector contains proportionally more of the radii. The sum of their lengths, although infinite, he took to be proportional to the area.

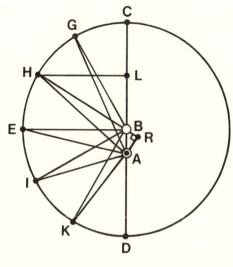

Fig. 16

Of much more interest to Kepler were the distances from the sun, at A in the figure. These distances, according to his physical analysis, measured the delays of the planet in the corresponding arcs of the orbit. They also were contained in sectors of the circle, with vertices at the sun, sectors such as CAG, GAH. Could he not measure the sum of all the distances to a section of the orbit by taking the area of the sector that covered them? Kepler thought so, not because the infinite sum could actually be formed, but because he believed that "in this area was a measure of the collected faculty which the distances possess for accumulating the delays."[79] In the finite approximation it seemed clear enough that the areas of triangles with equal bases were proportional to their altitudes. His equal divisions of the orbit were the equal bases. Would not the "triangle" erected upon one of these small arcs be proportional to its distance from the sun, and hence to the delay in that arc?

If this proportion were valid, he had found a solution in geometry to the problems of his physics. At any point of the orbit, say G in Figure 16, the mean anomaly from aphelion would be measured by a sector CAG with vertex at the sun, and the eccentric anomaly by a sector CBG with vertex at the center. The physical equation of center was by definition the difference between these quantities, namely the triangle GAB. (The optical equation, conveniently, was an angle AGB of this same triangle.) Kepler saw a fallacy in his reasoning, however, which forced him to conclude that the area law was only an approximation, as he was using it.[80]

---

[79] G. W., 3: 265: 1–2.
[80] Much confusion has surrounded this point. The clearest and most reliable discussions are those of Aiton, particularly in "Infinitesimals," pp. 295–6 and n. 38.

The problem was that when Kepler divided the orbit by radii from the sun, instead of the center, the radii were no longer perpendicular to the planet's path. Therefore it was no longer true that the sectors or "triangles" based on equal arcs of the path had areas proportional to their distances from the sun; for these distances were no longer the altitudes of the triangles. Being oblique to the bases, they were longer than the actual altitudes. Indeed, Kepler had no difficulty in showing that their sum was greater than the sum of the distances from the center. The latter sum was correctly measured by the area, so the former could not be. The method of computing equations from areas was simple, and had been developed from the physical causes of the equations rather than being a fictitious hypothesis; but it erred in measuring these physical causes by an area which imperfectly represented their effect. (It erred also, Kepler noted when revising this chapter for publication, in assuming the orbit to be circular. By that time he knew that the two errors cancelled. These same areas, although they did not measure the sum of the distances "contained" in them, precisely measured the physically-important sum of distances to the correct orbit.) Kepler did not understand these matters very well when he wrote the original draft of Chapter 40. They will come up again, on pp. 108–109 below.

What he did understand was the imperfection of the area law as he was using it. The difference was negligible for an orbit so little eccentric as the earth's, but he was about to resume the study of Mars. He needed some idea of the discrepancy if he was going to transfer his physical theory to that planet. For his analysis of the sums of distances, Kepler devised a curious figure, which for clarity we present as the three Figures 17a, b, c. Figure 17a represents an unrolling of half the circle of Figure 16, so that the circumference CGHEIKD is straightened, and the radii from the center, CB, GB, and the rest, are made parallel. The center B becomes a line segment closing off all the radii. Each

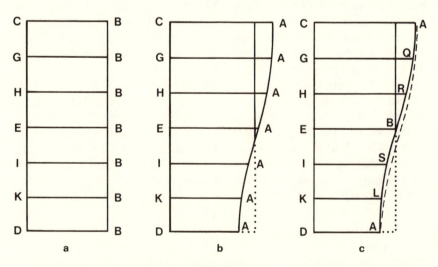

Fig. 17

sector or "Archimedean triangle," [81] such as CGB in Figure 16, has become a rectangle CGBB, with twice the area of the original sector. The area of the large rectangle CDBB is therefore twice that of the semicircle; and to any sector there corresponds a rectangle with exactly twice the area. The figure effectively displayed the proportionality between the area of a sector and the aggregate of the "distances" from the center B that were contained in it. (Kepler frequently used the word "distance," *distantia*, in a concrete sense to refer to an actual line segment, as here the line segment connecting the sun and the orbit. The usage was convenient, because of the importance of the distances as stand-ins for the delays, and we shall employ it often in the following pages.)

Distances from the sun were more complicated. As before, Kepler laid out the distances CA, GA, etc., parallel to one another. A curve AAAAAAA was required to close them this time (Figure 17b). In Figure 17 any section of the left boundary, such as GH, always represents the same arc of the circumference as in Figure 16. The area to the right of GH always contains, and represents the sum of, all the distances to that arc, whether from the center B as in Figure 17a or from the sun A as in Figure 17b. The area is *not* constructed to represent areas of circular sectors, and we must resist the hindsight which urges us to impose that signification upon it. It is constructed simply as an equivalent to the distances, by rotating them through whatever angle is needed to make them all perpendicular to the straightened circumference. With this done, they pave the areas of Figure 17 densely and uniformly, and are obviously equivalent to those areas.[82] In particular, the area of any horizontal slice from Figure 17b contains the distances from the sun to the arc represented by the left boundary of the slice, all laid out parallel, side by side. It is equivalent to the sum of these distances, and hence, by Kepler's fundamental law of physics, to the delay of the planet in traversing that arc of the circle. The problem for him was to measure such an area.

It was at this point, and only at this point, that areas of circular sectors entered the analysis. The distances from the center, used to construct Figure 17a, were constant, and clearly any slice from that figure was a rectangle whose area was proportional to the arc at its base (on the left). Measuring the area of Figure 17b, and of its horizontal slices, was much harder. The ends of the solar distances composing the area were closed off by an asymmetric curve Kepler called a conchoid. After a quadrant of the orbit (hence halfway down the figure, at E) the solar distance was still greater than the orbital radius, as is obvious in Figure 16. Overall, then, the conchoid seemed to delimit a greater area than the line BB.[83]

---

[81] G. W., 3: 268: 20–21.

[82] Whether Kepler could have "proved" this equivalence depends upon one's standards of proof, and is an irrelevant question unless one is studying the development of mathematical rigor. The equivalence is, and was, intuitively evident, and Kepler used it willingly as a tool.

[83] Kepler drew this conchoid, and drew from it the conclusion that the area law was an inexact measure of the distance law, in his letter to Fabricius dated February, 1605, #281 in G. W., 15: 30–31.

Note that one cannot compare the conchoid to the rectangle piecewise. The rectangle contains the distances to the center B, and is equivalent piecewise to sectorial areas from the center. The conchoid, on the other hand, contains the distances to the sun A. No sector on the circle has been found equivalent to them. They are *contained*, of course, in a sector with vertex at the sun, but as Kepler pointed out,[84] the distances do not intersect the circumference at right angles, so this sector is not properly equivalent to them. The excess or deficiency of a slice of the conchoid-figure with respect to the rectangular slice represents the physical equation; but it corresponds to no sum of partial distances, since the figures are composed from entirely different distances. Still less does it correspond to any area that can be delimited on Figure 16, the orbit.

The overall area to the conchoid seemed to be greater than the overall area to the line BB, judging from the respective distances EA, EB at quadrature. This made sense physically, for most of the circular orbit ("most" in terms of arc length) was farther from the sun than the mean distance or radius. Most of the delays were therefore greater than the mean delay that would obtain if the same size orbit were concentric to the sun. Making the orbit eccentric moved more of it away from the sun than toward the sun, and hence increased the periodic time. Kepler had demonstrated this in proving that the sum of distances from the center to equal divisions of the circumference was less than the sum from any other place. Alone, the excess of the conchoid area over the rectangle merely signified this increase in periodic time.

A much more serious difficulty in using the conchoid area was the lack of any equivalence between slices of it and useable areas of any kind on the planet's orbit, Figure 16. As noted several times above, the solar distances do not in general intersect the circle at right angles, so that one cannot view a circular sector as a triangle having a small division of arc as its base and the solar distance as its altitude. If one wants to measure sums of distances by sectors with a common vertex at the sun, so that the sum for any portion of the orbit can be represented by the corresponding sector, one should take for the distances the altitudes proper, measured perpendicularly from the bases. When the base lies on the circumference of a circle, the altitude is directed toward the center, along a line with no physical significance.

To meet the difficulty of measurement, Kepler constructed a second conchoid, shown as AQRBSLA in Figure 17c, from the distances which legitimately could serve as altitudes for sector-triangles. That is, to point H on the circumference in Figure 16 or 17 he constructed a distance equal to HR, where R is found by dropping a perpendicular from the sun to HB, the diameter of the circle which passes through H. HR thus really *was* the altitude of a triangle whose base was a tiny arc at H, and whose vertex was at the sun. In Chapter 40 this second conchoid was purely a measuring device, to which Kepler could compare the physically determined conchoid of Figure 17b. It was of no interest in itself.

---

[84] *G. W.*, 3: 267: 6–9; discussed above.

Qualitative comparison was easy. The second conchoid crossed line BB after exactly a quadrant, since the perpendicular distance of the sun from an arc at E is just the radius EB of the circle. Moreover, it is easily shown that the second conchoid is symmetric: however far outside of line BB it is at some point above the center E, it is the same amount inside line BB at the point just so far below the center. The wedge-shaped excess of the conchoid area over the rectangle, above the center, is thus congruent to the wedge-shaped deficiency below the center. The area bounded by the conchoid therefore precisely equals the rectangular area. Furthermore, the second conchoid was always inside, or to the left of, the first. (This is evident from Figure 16: a distance such as HR is always one leg of a right triangle whose hypotenuse is the solar distance HA.) This fact confirms, incidentally, that the original conchoid cut off an area greater than that of the rectangle.

Kepler noted that the area between the conchoids, while widest about halfway down the figure, was not symmetrical with respect to the midline EA. Rather it was wider below, near "perihelion" D. Although he did not remark the fact, this irregular variation mirrored the perplexing pattern he had studied with the epicyclic diagram reproduced in Figure 15b. On that figure, the irregular components of the distance increments, such as ικ and λμ, corresponded by construction to the width of the space between the conchoids, at the given eccentric anomaly.

At this point Kepler had a precise geometrical measure of the error introduced by using the area law to represent the delay of the planet in some part of the orbit. Physical theory taught him that the delay in any arc of the circle was proportional to the aggregate of all the distances to that arc from the sun, and hence to a slice of his first conchoid figure (Figure 17b). The area law yielded an answer proportional to the same slice taken from the second conchoid figure (Figure 17c). The difference, or error, for any part of the orbit could be computed if only he could determine the area of arbitrary horizontal slices of the tiny space between the two conchoids.[85] If this calculation had been possible, the resulting areas would have been normed by setting half of the planet's periodic time equal, not to the area of the rectangle, but to the area of the first, asymmetric conchoid, for that larger conchoid was the one which represented the physics of the motion. To his regret, he was unable to compute the area. He called upon the geometers of his time "to teach [him] how to square the space between the conchoids."[86]

The remarkable thing about these passages in Chapter 40 is that Kepler later discovered that the planet moved on an ellipse, in such a way that its distance from the sun, at a point corresponding—in a sense to be discussed— to any place on the circular circumference, was exactly that used in constructing the *second* conchoid. The area law was exactly in agreement with Kepler's

---

[85] Again, one must not imagine that this little space itself represents a sum of distances, nor that it is equivalent to some area in the circular diagram. It is the difference between the physically meaningful area in Figure 17b and another area, meaningless at present, which could be measured because of its equivalence to a circular sector.

[86] *G. W.*, 3: 269: 24–25.

physics; believing the orbit to be circular when he initially wrote Chapter 40, he had failed to recognize this agreement. The appearance of these precisely correct distances at this early stage in his work is highly suspicious.

Does it not appear, in fact, as if the second conchoid must have been added in a later revision of the chapter? Once Kepler knew the true distances to the ellipse, he would have found it almost irresistable to rework his murky and inconclusive analysis of the asymmetric conchoid, planting seeds of the wondrously simple solution which would emerge many chapters ahead. Here he introduced the second conchoid quite abruptly:

> However, EA [Figure 17b] is longer than EB. But if CA, GQ, HR, EB, IS, KL, DA were taken [Figure 17c now], as long as the perpendiculars determine which are dropped from A to the distances of the points from B ... then the figure between the conchoid AQRBSLA and CD would be exactly equal to figure CBBD.[87]

That is all the motivation we get. An odd construction gives an area which is equal, overall, to the rectangle of Figure 17a. Kepler did not even bother to show the piecewise equivalence, for any interval of arc on the circumference, of the area between the curves to the error of the area law. Little has been said about these questions in the secondary literature; scholars have generally ignored the entire analysis of the conchoid figures.[88]

How interesting it would be, though, to date the symmetric conchoid with Kepler's original draft of Chapter 40, well before he discovered that the distances from which it was constructed were correct. As we have seen, the shortened distances can be derived easily from Kepler's remarks on the area law: areas do not represent solar distances because the latter do not intersect the circumference at right angles, and hence cannot serve as altitudes for the "Archimedean triangles." If we take the actual altitudes of these triangles we get the distances used in Kepler's second conchoid. If in fact Kepler's early analysis of the area law did lead him to the symmetric conchoid in this way, we would have the surprising conclusion that Kepler first employed the correct distances (to the ellipse) while trying to evaluate the error of the area law applied to the circle. At that time, of course, he had no inkling of any physical significance that could attach to the shortened distances; nor of the relation, far from obvious, between the distances stretched across Figure 17c and the arc length measured down that figure. But the distances would have been there, in his draft chapter, awaiting the moment years later when he could make those links.

I think it probable that he did construct the second conchoid, ignorant of its significance, in just this way. His published correspondence is of no help

---

[87] *G. W.*, 3: 268: 36 – 269: 2.

[88] Caspar noticed the figures, and remarked that Kepler "braucht diese Überlegungen im folgenden nicht mehr, hat aber seine Freude an ihnen." *Neue Astronomie* (Munich: R. Oldenbourg, 1929), p. 406; also *G. W.*, 3: 470.

in dating these portions of Chapter 40. Consider, however, the way in which he closed his comparison of the conchoids, by calling for the aid of geometers to help him "square the space between the conchoids." After Kepler had found the ellipse, after he knew why the second of his conchoids was so interesting, this problem was moot. The tiny space measured nothing of any consequence. The sum of distances corresponding to the first conchoid—which was the problem here, for the second was measured precisely by the sector of a circle—was no longer of any concern to Kepler. When he called for the help of other mathematicians he surely thought the problem important; and if so, he did not yet know that the easily measured area was the one he wanted. On this morsel of evidence, slight but persuasive, I think we can conclude that Kepler used the relation between eccentric anomaly and distance, while testing his area-law approximation, several years before he realized that both it and the area law described precisely the motion of a planet.

This fortieth chapter concluded Part III of the *Astronomia nova*, a section he had called the key to a more profound astronomy. Exploiting the precision of Tycho's observations and of his own vicarious hypothesis, Kepler had charted the irregularities of the earth's motion, in distance as well as in longitude, by analyzing the second anomaly of the motion of Mars. To accompany these important technical refinements he had sketched, boldly if naively, an utterly new field of inquiry for the astronomer. The physics of planetary motion, the pattern according to which that motion was guided, he now announced as the way to an exact knowledge of the celestial motions. Kepler had forsaken the elegant techniques of his predecessors for doubtful ratios of infinite sums. His investigations of physical causes had posed new problems while offering hints at the solution to the older problems, and had removed the harmonies Kepler sought to a level beneath the appearances. The vindication of this new kind of astronomy lay yet in the future, though, for the old methods, ingeniously applied, had been adequate for the theory of the earth. It was Mars, most obstinate of the ancient planets, which would test the powers of his physical astronomy, and it was Mars which would reveal whether a new elegance was to be found in physical forces behind the appearances in the heavens.

Conquest of Mars

As a preliminary to his main attack on Mars, Kepler took the three positions of Mars from which he had measured the earth's orbit in Chapters 26–28, and calculated the circle on which they lay. In passing, he advised his reader that other positions of Mars would yield different circles. And in fact, the eccentricity he found here in Chapter 41 was not at all that obtained when he had analyzed the equations of center and of latitude in Chapters 16–19. He advised his readers that analysis in the very next chapter was going to confirm that earlier value for the eccentricity, rather than this new one based upon three positions.

The positions were reliable; evidently, then, the circle they defined was not the orbit.

To determine more closely the eccentricity and apsidal line of the orbit, Kepler selected in Chapter 42 a group of observations separated by integral revolutions of Mars, with that planet near aphelion, and a similar group with Mars near perihelion. For each observation Kepler knew the distance and direction from the earth to the sun, from the table in Chapter 30, and the direction to Mars, from observation. If he assumed a value for the distance between Mars and the sun, the triangle sun-earth-Mars was given in shape and position. In particular, the heliocentric longitude of Mars was determined. Kepler's procedure was simply to adjust the assumed distance of Mars from the sun until this longitude came out the same for each one of the group of observations. As a further check he calculated the aphelial distance with a procedure he had developed in Chapter 28. (There he had used it indirectly, to confirm an assumed value for the bisected eccentricity of the earth, by requiring that the distance between Mars and the sun come out the same, for different groups of observations with Mars at the same place. Here he calculated directly the distance of Mars using the known eccentricity of the earth.)

Kepler now had two reliable positions of the planet near aphelion and perihelion. Comparing the time intervals and longitudes, he was able to show that the planet must have been about forty minutes of arc past aphelion at the former point. He based this adjustment on his knowledge that near the apsides the real motion was inversely proportional to distance from the sun, so that apparent motion in longitude was inversely proportional to the square of the distance. His new value for the apsidal line, based upon distances observed from the earth, agreed with that computed from the equations of center in Chapter 16, after corrections for precession and the slow motion of the apsides.

Furthermore, he now had values for the greatest and least distances of Mars from the sun based directly upon observation, and not upon any particular hypothesis about that planet's motion. The only specific assumption he had made about Mars was its periodic time, which he knew reliably from comparison with ancient observations. (The assumption about apsidal velocities mentioned in the last paragraph was made to estimate angular corrections of about a minute; at the apsides, these obviously had no effect upon distance.) From the greatest and least distance he knew the eccentricity of the orbit. To nobody's surprise who had read this far, it was half the eccentricity of the Martian equant. "So that in Mars also," he reminded the reader in his summary of Chapter 42, "the preceding speculations from Chapter 32 on are valid." [89]

He proceeded to test them again in Chapter 43. At a quadrant from aphelion (point E in Figure 16), the elapsed time was proportional to the sum of distances from the sun to all the points in the arc CE which Mars had

---

[89] *G. W.*, 3: 46: 36–37.

traversed. This sum was approximately proportional to the area of the sector ACE, by the reasoning of Chapter 40. The physical equation, or deviation from uniformity, therefore roughly equaled triangle ABE, in the same proportion as that of the periodic time to the area of the circle. At other places in the orbit, such as H, the equation corresponded to a triangle ABH, which had the same base AB as the previous one. Thus the physical equation at H equaled its greatest value ABE scaled down by the ratio of the altitudes, HL to EB, which was simply the sine of the eccentric anomaly CBH. The optical equations he could find (as before) by solving for the vertex angles of these same triangles.

We do not know at how many points Kepler checked the equations that followed from the area law. He published three comparisons, for these were quite enough to show the defectiveness of his methods. The comparisons, of course, were not directly with observation, but with the equations from his vicarious hypothesis of Chapter 16. At the quadrant, with eccentric anomaly $\beta = 90°$, the new calculation (with the area law approximation) was very accurate. At the octants, $\beta = 45°$ and $135°$, it was in error by about eight minutes. The planet, as he put it, was further from aphelion than it should be in the superior quadrant ($\beta = 45°$), and further from perihelion than it should be in the inferior quadrant.

These locutions are not entirely clear. When Kepler evaluated this theory by saying that Mars was too far from aphelion at the upper octant, he was criticizing not the planet Mars but the theoretical calculation of Mars's position. His theory predicted a position too far from aphelion, compared to a known value. (Of course, the known value here was itself a theoretical prediction, and moreover one from an untenable theory. Kepler knew, however, that the vicarious hypothesis gave equations accurate to a couple of minutes.) We shall follow this sensible convention. Rather than saying that the theoretical value which we are presently considering for a planet's longitude or distance was greater than some value known to be accurate, we shall simply say that the longitude or distance of the planet was too great— meaning thereby the theoretical planet. Under this convention, then, Kepler concluded that the planet was too fast near the apsides, and correspondingly too slow in the mean distances. He needed a theory that would make the planet move faster around the quadrants.

Kepler determined, before proceeding, that his method of approximation— namely the area law—was not responsible for the error. He actually attempted, by assuming or guessing a mathematical form for the width of the area between the conchoids,[90] to calculate the area between his two conchoids. The calculation, not very accurate and completely useless from his vantage point of a few years later, lends more support to our suggestion (above, p. 86 f.) that Kepler worked with the symmetric conchoid, and hence with the correct

---

[90] He supposed this width to vary about as $\sin^2 \beta$; see Caspar's discussion in G. W., 3: 471, or in Neue Astronomie, p. 407.

functional form for distances, at a very early stage in his investigations of the distance law. (We say the correct *functional* form because Kepler was still attempting to use the traditional definition of the eccentric anomaly β in evaluating these distances. He was not using the correct distances, but he was using the correct equation, namely R = 1 + e · cos β.)

These computations were hardly necessary, though, for as Kepler pointed out,[91] the area law erred in the wrong direction to account for the empirical failure of his theory. The distances equivalent to the area law were too short to represent the physical distances, but they fell short most especially in the mean distances. More accurate computation would use longer distances, hence longer delays, in the mean distances, aggravating the error in the theory. The problem, he concluded, lay elsewhere than in the area law.

It took Kepler two years to accept the solution to his dilemma. Once accepted it was rather obvious. The orbit of Mars was not a circle; it was an oval. Direct trigonometric calculations of the circle passing through any three observed points, of the kind exemplified in Chapter 41, yielded different circles for different triplets of observed points. The circle of preference was the one passing through aphelion and perihelion, points which Kepler had carefully located in Chapter 42. However, distances to other points of the orbit fell short of this circle.[92] The amount of shortfall varied, being greatest in the quadrants. Compared to the eccentric circle, the orbit of Mars was squeezed in at the sides.

Kepler's physics confirmed this, for the equations he had computed placed the planet too far from the apsidal line. If the planet actually was closer to the sun than he had thought in the mean distances, it was moved faster by the solar virtue in those regions. This theory, using the distance law on a circular orbit, moved Mars too slowly in the mean distances, and hence too swiftly near the apsides. At an octant past aphelion, the theory had moved Mars too far, just as his distance calculations implied; and at an octant before perihelion the theory had not moved Mars far enough, which again he had found to be true. The distance calculations were imprecise, and the physics still speculative. Nonetheless, the two arguments converged on the same conclusion: that the orbit was an oval of some kind.

We can easily acknowledge the extraordinary attraction of circles in ancient and medieval astronomy. Most of the reasons for astronomers' loyalty to circular models are probably well known. The circle is the simplest closed curve. It revolves readily. It gave aesthetic pleasure to Pythagoras, we are told, to Plato for certain, and to others after them. It represents the motion of a point on a rotating sphere. One can calculate with trigonometry on a circle. Astronomers had always used circles. We can talk about these reasons today, but really we are in no position to appreciate their force or to compare their

---

[91] *G. W.*, 3: 283: 38 – 284: 3.

[92] Kepler reached this conclusion in 1603, while performing the analysis of the earth's orbit reported in Chapter 26: see the letter to Fabricius, #262, in *G. W.*, 14: 410–411. Contrary to one's first impression, the *Astronomia nova* is arranged not chronologically, but rather for didactic reasons.

significance, as they appeared to Kepler and his contemporaries. Individuals valued them, no doubt, in differing proportion. Kepler was one of the first astronomers to be free of perhaps the most objective basis for circularity, the assumption of solid spheres. Though his was one of the most boldly original minds ever devoted to astronomy, he hesitated long before abandoning the circle. Upon finally taking that step, he found himself no longer on the verge of attaining his goal, but instead adrift in a sea of new questions.

In violating the tradition of circularity, Kepler had to face the full import of one of the lesser supports we have cited for that tradition. Circles were easy to calculate with. In part this was merely one aspect of the more general difficulty: Kepler had replaced a specific hypothesis with a vague one. A circle was a circle, but he had no idea what his oval was. Let us stay with the specific problem of calculation, however, for that specifically was how the new difficulties presented themselves. Motions composed from uniformly moving circles, such as the old astronomy had employed, submit readily to trigonometry. Kepler already had discarded the mechanism of uniformly rotating circles, thereby bringing upon himself the computational nightmare of his distance law. He had thus far retained (in fact, insisted upon) circularity in the shape of the orbit, however, and had very nearly reduced his infinity of distances to an easily measured sector of the area of that circle. Now he had tried to locate the circle, and had been frustrated.

The new theory with which he replaced the failed one did not, of course, have to be composed of circles; Kepler had gotten past that requirement in his concept of the physical world. It did have to be a precise theory, though, one which mathematically defined the movement of the planet around the sun. He had eliminated the obvious candidate for the shape of the orbit, and it was by no means evident where he would find another that would be both physically accurate and mathematically tractable. Indeed, it was no longer clear whether these two goals were even compatible.

At the end of Chapter 39 (p. 79, above), Kepler had been unable to understand how it was physically possible for a planet to follow an eccentric circular path, and had laid aside that problem while he developed a way to use his distance law. His discovery that the orbit was not circular had gotten him out of the impasse in his physics, so he returned quickly, too quickly, to the physical arguments of that earlier chapter. To recapitulate them briefly: each planet somehow approached the sun in passing from aphelion to perihelion. In the circular hypothesis one could conveniently represent its approach with a diagram employing an epicycle (Figure 15b), although the epicycle was not real, but merely a geometrical construction used to describe the action of the physical forces involved. The difficulty was that, in order to keep the orbit circular, the epicyclic model had to turn at the rate of the planet's eccentric anomaly. This was irregular motion indeed, and simplified not at all the task of understanding change in distance from the sun. Eccentric circular motion seemed to admit no plausible physical model, once solid spheres had been abandoned.

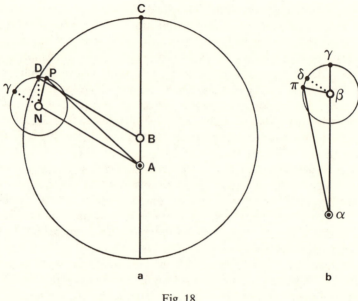

Fig. 18

An oval orbit, constricted slightly around the mean distances, posed no such problem. If one described the change in distance by *uniform* motion on an epicycle, one obtained a constricted orbit. In Figure 18b, let the planet's distance απ be determined by uniform rotation of the epicycle. As the planet departed from its greatest distance at aphelion γ, its eccentric anomaly lagged behind the time, or mean anomaly. In the circular model one set angle γβδ equal to the eccentric anomaly, and obtained the distance as αδ. Here angle γβπ was the mean anomaly, greater than the former angle, so the distance corresponding to the same eccentric anomaly was απ, a little less than before. Transferred to a representation of the orbit (Figure 18a), this diminution of the distance brought the planet inside its former circular orbit. At perihelion F the planet again lay on the circle. Kepler proposed this oval in Chapter 45.[93]

This first oval model reminds us forcibly of the poverty of physical theory, and of mathematics suitable for physical analysis, available to Kepler. Chapter 45 contained no suggestions about the reason for the pattern of motion being described. For these the reader was referred back to Chapter 39, which had simply asserted that there must be a virtue proper to each planet. Yet Kepler repeatedly described the model of Chapter 45 as "physical." As a physical model it certainly seems a step backwards. In the second chapter of the book Kepler had argued against the reality of epicycles, yet here he built a physical theory upon one.

---

[93] He was using the uniformly-rotating epicycle as a model for the oval by October of 1602, as shown by the letter to Fabricius, #226 in *G. W.*, 14: 277–278.

In fact, Kepler was not recanting his opinions. He did not attribute physical reality to this epicycle, but used it because he had to use some device to construct a theory. He needed a geometrically precise way to specify where the planet was. He used an epicycle, more or less as we would use a trigonometric function, to represent a periodically varying distance. So long as the epicycle had to turn nonuniformly, it had been an unsatisfactory device, and he had rejected it for this reason in Chapter 39. After he discovered that the orbit was oval, he could use a uniformly turning epicycle, which was workable as a theoretical device. This epicycle then became the principal candidate for study. Through all of this Kepler's "physics" involved very little positive physical theory. It was, instead, a sort of analytical viewpoint, from which the patterns of change in physical quantities—distance and motion, above all—were seen as clues to the source and working of the physical causes. The causes themselves remained mysterious, although Kepler enjoyed speculating about them; but knowledge of *how* they functioned was indispensable to the new astronomy. This kind of analysis had comprised the critical physics of Chapter 39. From it sprang the distance theory of Chapter 45, a theory which was physical not because of the putative epicycle it described, but because of the pattern of changing distances which that epicycle represented.

This is not to say that Kepler was consistent in his relegation of the epicycle to the mathematical toolbox. There are places in his writings which seem to show concern about the consequences of a real uniformly-rotating epicycle. In a long letter to Fabricius dated October 1, 1602,[94] Kepler discussed, among many other things, this first oval model. He characterized the motion of Mars itself as one of "struggling against the virtue of the sun," and moreover of being "directly backwards" at aphelion.[95] This describes the motion of a point on an epicycle, not of the planet. Later in the same letter Kepler speculated whether it would not be better to have the planet move on an epicycle perpendicular to an "imaginary mean orbit," so that the epicyclic component of motion would not interfere with the solar virtue.[96] We must not be misled, however. Kepler may not have been clear just what combination of physical motions he was proposing, but he knew very well that the epicyclic motion was not one of them. "When I say that the planet regards the virtue of the sun in struggling, I am explaining the physical cause of the planet's motion. When I say it regards the center B [of the epicycle] I am explaining the mode of our understanding." [97] The epicycle was an aid in describing the motion, but it

---

[94] *G. W.*, 14: 263–280. The section of interest here is from 277: 563–279: 615. The figure on p. 278 is not consistent with the text, which sometimes (but not always) reads as if the labels should be reversed for points C and E. The diagram in Frisch's edition has the two labels reversed, and his text is always consistent with his diagram. *Joannis Kepleri Opera Omnia* (Frankfurt: Heyder & Zimmer, 1858–1891), 3: 68 (hereafter cited as *O.O.*).

[95] "Is motus Martis consistit in nitendo contra virtutem Solis. Nam in principio temporis restitutorii seu anomaliae nititur directè retro...." *G. W.*, 14: 277: 567–568.

[96] *G. W.*, 14: 279: 628–645.

[97] *G. W.*, 14: 277: 572–575.

had nothing to do with the physical forces at work. Kepler's discussions of the epicycle were to clarify its implications as a mathematical model, not as a physical model.

For the still-rudimentary physical model we must fix our attention on the radius AP connecting the sun and the planet in Figure 18a. The rotating solar image carried the planet around the sun, with delays proportional to distance AP in a given small arc of motion. Meanwhile the planet, by some innate virtue, moved toward and away from the sun *along AP* in a manner that changed the distances just like uniform motion on an epicycle. The epicyclic center N, and the constant radius AN connecting it with the sun, were mere constructions used in determining the length of AP. It is helpful to conceive of N as an imaginary point determined by the constant distances AN and PN, plus the requirement that angle PNA change uniformly. This imaginary point, the epicyclic center, moved around the sun on a circle (by construction), very nearly as the eccentric anomaly.

Actually, one could no longer say what should be termed the eccentric anomaly. Until Kepler had abandoned circular orbits, eccentric anomaly had been the fundamental physical parameter of the planet's motion. It was the arc between aphelion and the planet's position (or, equivalently, the angle between aphelion and the planet, measured at the center of the circle). As such it directly traced out the motion of the planet. Kepler had formulated his distance law with respect to small equal increments of eccentric anomaly, that is, of distance covered. With a noncircular orbit, however, he could no longer use the eccentric anomaly indifferently as both an arc of the orbit and a central angle; and so, in the chapters from 45 until 58, the eccentric anomaly remained in a sort of limbo. It could not be dispensed with, for it was an essential component of the physics, but it was not clear just how it should be combined with the geometrical description of the orbit. Kepler preferred to regard eccentric anomaly as arc length on the oval orbit, since that was the way it entered into his physics; yet he was unable to measure the arc length on a noncircular orbit.

This problem appeared first when Kepler attempted, in Chapter 46, to combine his epicyclic model for the planet's distances from the sun with his distance law for motion around the sun. The former theory determined distance as a function of time, or mean anomaly, while the latter theory computed the time as a function of eccentric anomaly. Since Kepler lacked any adequate way of representing eccentric anomaly, the combination of the two theories was far from easy. Without combining them, however, he could not determine a planetary position.

Kepler's analysis of the orbit ensuing from the distance model of Chapter 45, in conjunction with the distance law for eccentric anomaly, is noteworthy in that he analyzed the motion into equal increments of time, rather than distance. For the first time, therefore, he restated his distance law as an inverse proportionality of path traversed to distance from sun, for equal increments of time. Previously, he had always written of the direct proportionality between

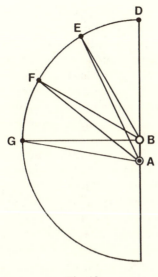

Fig. 19

delay and distance from the sun. Now he divided the orbit into small pieces traversed in equal times, and stated that the ratio of the length of any one piece to the length of another was inversely as the distance from the sun of that first piece to the second.[98]

Kepler knew the distance from any point in the orbit to the sun, more or less, and he knew the average distance from the orbit to the sun. The ratio of these two distances was proportional to the motion of the planet at that point in the orbit. If he could accumulate these ratios of distances, at each increment of time, he would have the sum of the increments of motion, and thus the total distance the planet traveled, during any given interval of time. He had no way of summing the ratios. He did, however, know how to sum the numerators and denominators of the ratio, for he learned in Chapter 40 (pp. 81–85) that sums of distances were very nearly equivalent to areas.

Notice that the distances in question here were not the same as those he had summed in that earlier chapter. He was using the technique, not the results, of his earlier work. There he had taken distances corresponding to equal path elements, and tried to calculate the unequal times required to traverse them. Here he took distances at equal increments of time, and attempted to calculate the unequal path elements corresponding to these increments of time.

To sum the distances, Kepler transformed the epicycle of his distance model into an eccentric (Figure 19). Like the epicycle, this eccentric was purely a geometrical construction. He termed it "fictitious," for it did not represent the

[98] *G. W.*, 3: 292: 17–19.

orbit of the planet at all. Rather, it was a display of the distances derived from the epicyclic model, arranged to allow him to construct an area containing them all, regularly spaced; an area therefore that was equivalent, or nearly so, to their sum. The radius of this eccentric was that of the orbit, equal to αβ in the epicyclic model of Figure 18b. The "eccentricity" AB was the radius of the epicycle. Therefore, by the basic equivalence theorem for epicycles and eccentrics, the distance from A to any point of the eccentric, say F, was equal to the distance from α to a point π on the epicyclic representation, providing only that the angle DBF, between aphelion and point F, equaled the angle γβπ between aphelion and point π. *Neither* diagram contained any information about the location of the planet; only distances were involved. The angles at the center B of the eccentric measured time, or mean anomaly, because of their equivalence to the angles in the uniformly rotating epicycle. Moreover— and this was the point of the eccentric transformation—a sector of the eccentric with vertex at A was almost equivalent to the sum of all the distances from A to equally-spaced points on the arc delimiting that sector. Kepler had obtained this approximate result, on an eccentric which did represent the orbit, in Chapter 40. Here he applied it to the fictitious eccentric.

The same diagram could trivially be used to sum the mean distance for the same number of time intervals: one merely took a sector from the center of the eccentric. With these two areas Kepler reasoned as follows. The element of its path which the planet actually traversed was to the element it would have traversed at mean velocity as its mean distance was to its actual distance. Summing, and converting to areas, the distance traveled by the planet was to the distance it would have traveled at mean velocity as the sector from B was to the sector from A, on the fictitious eccentric. Together with the planet's distance from the sun, which was taken directly from Figure 19, this proportionality seemed to determine the planetary position.

Ingenious as it was, Kepler's construction was marred by a number of inaccuracies, which he immediately enumerated.[99] First of all, sectorial areas represented only imperfectly the sum of the distances contained in them. More seriously, he had taken a proportionality between increments of path and applied it improperly to the sums of the increments. He had little alternative, since he lacked even an approximate technique to average the ratios of the distances. The practice was inexact, but he argued that his error was small in this instance. (It is clear that Kepler intended a genuinely physical construction based on very small elements, as we shall see on p. 98. The other flaws of the technique perhaps precluded a more elaborate discussion at this point, such as would have been required for an element-by-element construction.) A further difficulty was that although the proportion determined the length of the desired section of orbit, it did not provide a geometrical construction of this length. Finally, the orbit was an oval. Even if it had been possible to construct a circular arc of the correct length, the planet's path would be inside it, and could not be located by any direct construction.

---

[99] *G. W.*, 3: 292: 27–293: 20.

Kepler was able to modify his technique to avoid the difficulties in applying the inverse-distance proportionality to a sum. When constructing the planet's distance on the fictitious eccentric, he had used a central angle equal to the mean anomaly, as discussed above. The equivalent area from the center delimited an area equal to the mean anomaly. By marking off on the real eccentric a sector from the sun of area equal to this, and measuring out the correct distance from the sun, he escaped from all his distance summations, and attendant problems. Alas, the other flaws remained. The area law was still not exact, the required construction was still impossible, and he still lacked any direct means of determining how tightly his oval should be constructed.

Kepler was finally driven to a construction which made no pretense of reflecting what was actually happening to the planet. His vicarious hypothesis gave him an accurate longitude; the equally fictitious epicycle introduced in Chapter 45 gave him a distance. Between them they produced, point by point, an orbit with good heliocentric longitudes and an oval shape. (The procedures here in Chapter 47 turn out to be significant for Kepler's final acceptance of the ellipse, as we shall argue on pp. 126 ff.)

This last method, unsatisfactory though it was, at least enabled him to use the epicyclic distances to determine some planetary positions. He set out to calculate the width of the very narrow slice or "lunula" by which the oval differed from a circle. The construction was laborious, and only approximately correct, but in the end it yielded the numbers he needed to proceed with the evaluation of his mathematical distance model.[100] The greatest width of the lunula was about .00858 of the orbital radius. To estimate its area he substituted an ellipse for his asymmetric oval, on the assumption that the two curves differed by little, and found that the lunula cut off by the ellipse had the same area as a circle of radius equal to the eccentricity of the eccentric circle. This was, in fact, about .00858 of the overall area of the circle.

To compute equations by his physical theory, or by the area-law approximation to it, Kepler required not only the total area of his oval, but the area of an arbitrary sector with vertex at the sun. This he could not obtain, for he had no idea how to measure the asymmetry of the oval. His only recourse, as before, was to ignore the asymmetry and calculate with sectors of the auxiliary ellipse. By this method he computed equations at 45°, 90°, and 135° of eccentric anomaly, and set them out in a table for comparison with his various other models. These were the area law on a circular orbit; equant models with bisected eccentricity and with the "vicarious" division; his terrestrial model with bisected eccentricity and assumed equality of the physical and optical equations; and a simple eccentric. Compared to the vicarious hypothesis, still his only standard, the new oval was the most accurate. It erred, at the octants,

---

[100] Incidentally, this construction in Chapter 47 did not employ the vicarious hypothesis itself for longitude, but instead a bisected-eccentricity equant model. He had found such a model inadequate in Chapter 19, as it erred by as much as eight minutes in the octants. Here, however, he wanted the position near a quadrant of eccentric anomaly, where the lunula was widest; and the bisected-eccentricity model had been accurate at the quadrants. Its use simplified a diagram which was yet none too simple.

by over six minutes—errors opposite to, and only slightly smaller than, those of the circular orbit with physically based equations. The circular orbit had moved the planet too fast near the apsides, and too slowly in the mean distances; this oval did the reverse.

Dissatisfied with the approximations he had been using (of which the most notable, for us, were the approximation of the orbit with an ellipse, and the approximation of the physical equation with the area law!), Kepler embarked upon lengthy calculations aimed at a more accurate representation of his theory. Six of these he preserved in Chapter 50. We need not work through all of the details. He struggled with his distance law, trying to find an interpolation that would yield a distance for each equal increment of the planet's path. Since his distance model was a function of time, and he did not know the shape of the path, he could not do this directly.

During the course of Chapter 50, Kepler tried a reformulation of the distance law: delays in equal increments of *angular* motion about the sun should be as the squares of the distances. This formulation was based on the idea that the path length in an angular increment increased linearly with distance, as did the delay in unit path length. It was not quite equivalent to the theory he had been using, for only the component of arc length perpendicular to the solar radius increases proportionally to distance from the sun. The component along that radius varies according to the shape of the orbit. Kepler was in fact trying a different distance law in Chapter 50, for the first time—one precisely equivalent to the area law.[101] Even so, to find the planet after an increment of coequated anomaly (in other words, an increment of angle around the sun), he was forced to fall back on the vicarious hypothesis. Direct calculation from his first principles eluded him.

All of his methods had failed at the moment when, by one device or another, he tried to transfer the distance from his epicyclic model to its proper place on the oval. He simply did not know where the oval was, and hence he was unable to place the distances correctly. In Chapter 48, disavowing all geometrical expedients, he set out to calculate planetary positions physically at whatever cost. For each degree of mean anomaly he determined the distance from his mathematical model of Chapter 45. According to his distance law— for the area law was now in disrepute—the ratio of each of these distances to their sum was inversely as the element of path traversed during the 360th part of the periodic time was to the total length of the oval path.

The circumference of the oval was essential to these calculations. Kepler first approximated it by an elaborate construction of a circle with circumference nearly equal to the oval,[102] but in the end he had to correct his value so

---

[101] This formulation has not, to my knowledge, been noted in the literature. It is deeply buried in the catacombs of Chapter 50, and even there Kepler did not remark upon the difference between it and his original formulation.

[102] Chapter 48 includes the construction and rationale for this approximation. In the construction figure prominently both the maximum width of the lunula, 858 parts of 100,000, and the half of that width, 429 parts. It is not at all surprising that these numbers stuck in Kepler's memory.

as to put the planet just at perihelion after half the periodic time, and then recompute all his proportions. From the solar distance, and the distance traveled during the first degree of mean anomaly, he had the position at the end of that degree. Thence he could proceed to the second degree, and so on through 180. The approximations of the area law, of the elliptical orbit, of the cumulated proportions, and of the amount by which the oval curved in from the circle were avoided. Only numerical and rounding error, and the approximation with one-degree arcs, remained; and he had rendered these largely noncumulative by requiring Mars to reach perihelion at the proper time.

The numerical error in Kepler's method is of some interest as it relates to the evolution of the eccentric anomaly. Finding the position of Mars, from its solar distance and the most recent increment of its motion, amounted to determining the coequated or true anomaly, the planet's angular distance from aphelion as measured at the sun. The path distance was quite an inconvenient parameter, measured as it was along a curve defined only numerically. Kepler used it in a roundabout way (Figure 20). Constructing lines AC and BC, from the sun and the orbital center respectively, through the imperfectly known

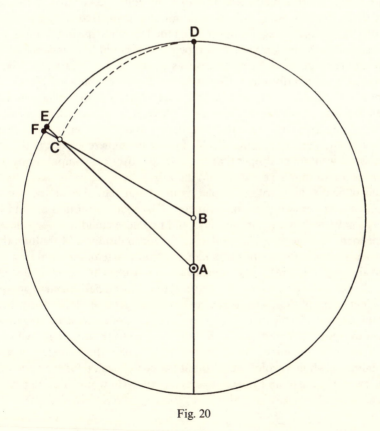

Fig. 20

planetary position C, he let them intersect the circumscribing circle at E and F, respectively. Because the orbit was so close to the circle, E and F could be considered coincident; the apparent distance between them was always less than a second of arc. Kepler wanted the coequated anomaly, angle DAC. Since he was given the solar distance AC and the eccentricity AB, the law of sines would give him the coequated anomaly from the other angle DBC, and hence from arc DF or its near-equal DE. These arcs, of course, he could find by adding increments of circular arc to the arcs calculated for the previous degree of mean anomaly; but the increment of arc on the circle was not given. Kepler had to estimate it by projecting the path increment out onto the circle, increasing it proportionally over the distance it was projected. Without going into the details, which were a bit messy, we note that Kepler was in effect converting, with considerable difficulty, between various numbers which had some claim to the name of eccentric anomaly: the central angle DBC, the equivalent circular arc DF, and the distance along the path, DC, which was so fundamental to his physics. He did not use the term "eccentric anomaly" at all in this chapter, although it had been the basis for all his analysis when he worked with a circular orbit.

After all this exceedingly tedious work, the equations in the octants still erred by as much as four and one-half minutes, compared to the vicarious hypothesis. The planet, again, was moving too fast in the mean distances, and not fast enough at the apsides. This type of error could have meant that his mathematical model for the distances was wrong, pulling Mars too close to the sun in the mean distances. Kepler, however, was encouraged by the way his more meticulous calculations had improved the equations, so he continued to develop that same theory. He was *not* pleased by the expedients to which he had been driven in calculating positions; and particularly he was not pleased by the need to assume a value for the circumference of the oval.[103] His distaste went much deeper than mere tedium over the computations. The means he found to relieve it were odd indeed.

The difficulty had arisen, Kepler pointed out, because of the mutual dependence of the two parts of his theory. The planet's motion around the sun was governed by the distance law, and therefore could not be evaluated independently of the epicyclic model for change in distance. More than that, the calculation of any one position required knowledge of the total length of the oval path, for the ratio of the planet's mean distance to its actual distance equaled that of the element of path traveled to the mean path element, in some small increment of time. The final term of this proportion blocked any direct solution, because the mean distance traveled during a degree of mean anomaly depended upon the length, and hence the shape, of the entire path; which is to say upon the combined working of both the solar and planetary virtues, throughout a whole revolution. Without the correct value for the circumference, even a painstaking degree-by-degree analysis would not return the

---

[103] *G. W.*, 3: 310: 23–31.

planet to its aphelion at the completion of an orbital period. This awkwardness was inherent in the physical character of the distance law, which operated
according to the remoteness of the planet itself, whatever the other influences
might be on the planetary position.

Kepler tried to evade these difficulties in a curious and revealing manner.
He modified his theory so that it was the imaginary epicyclic center, rather
than the planet, which moved in inverse proportion to the distance separating
planet and sun. Referring back to Figure 18a, we now say that N moves
around A inversely as the distance AP; meanwhile the angle γNP on the
epicycle increases uniformly with time, as before. (Equivalently, D moves
around the vacant center of the orbit, B, inversely as AP.) The new model is
not easily compared with that of Chapters 45–48, where the planet itself had
moved inversely as its solar distance. Inspection of the diagram shows that it
differs from the circular distance-law theory, of Chapters 39–43, by the small
circular arc DP, whose magnitude is the physical equation of center on the
circular model.

In any event, this new hypothesis immediately presented serious problems
of physical interpretation. How could the epicyclic center, an imaginary point
whose distance from the sun never varied, follow in its motion the proportion
of the distances between two physical bodies, neither of which moved in that
or any other regular manner? This oval was absurd, as Kepler confessed in a
note added later, for four of the five reasons adduced against its circular
predecessor in Chapter 39 (p. 78, above). He explicitly cautioned here that
the epicyclic center moved "not on its own account, since it is not a body, but
on account of the planet"[104]—a necessary reminder, since the description of
the model itself was laid out the other way around. Moreover, the new model
did not work. The equations, laboriously computed a degree at a time, as
before, were decidedly worse than they had been.

For what reason had Kepler done all this? His equations were bad, his
physics garbled. A small consolation, notable only because he mentioned
it himself, was the sharper separation in this new theory between the two
component models. He had hypothesized a solar virtue varying inversely with
distance in its effect, and a planetary virtue absolutely constant. Previously
he had tried to apply the distance law to the planetary position itself, the
resultant of both virtues' working. With his new model the two patterns of
motion were distinct, because the distance law applied to the center of the
epicycle whose rotation described the action of the planetary virtue. This
justification is of some interest, but is hardly adequate. Neither of the components, so artificially separated, was satisfactory. The solar virtue was no
longer of plausible simplicity. It acted on the planet, but the intelligible pattern
of its action hinged on an unreal point. In this new theory, the planetary virtue
had disturbed the distance law by extricating the planet from the *radius
virtuosus*.

---

[104] *G. W.*, 3: 312: 15–16.

This *radius virtuosus* (see above, p. 76 f.) was a fictitious construct which Kepler employed when discussing the interaction between the moving image of the sun and the planet's own action in approaching or withdrawing from the sun. It consisted of a moving radius along which the planet would lie if acted on by the solar virtue alone. This was an elusive concept, since the radius moved at a rate that depended on the planet's distance from the sun—the component from which Kepler was abstracting when he spoke of the *radius virtuosus*. The *radius* represented the longitude the planet would have at any moment, if nothing moved it in longitude except the rotating image of the sun. This virtuous radius "served as a place for the planet," that is, it served as a frame of reference for the planet's proper motion.[105] Any motion presupposed a locus, a place where the object would be were it not moved. In the presence of other forces this locus could itself move in a complex way; and indeed the locus of the *radius virtuosus* was swept around by the sun's moving image.

Previously Kepler had thought the planetary virtue to act directly toward and away from the sun, strictly within the *radius virtuosus*. The body of the planet obeyed the distance law, and could be thought to maintain the locus represented by the ray joining it to the sun. This locus carried the planet around the sun at a rate inversely proportional to the length of the radius. The planetary virtue, acting along the radius, had never carried its body out of this moving locus; not until, that is, the innovation of this forty-ninth chapter. Now the planet itself no longer followed the distance law. Evidently it extricated itself (the word is Kepler's) from the generalized locus of the *radius virtuosus*. The planetary motion was no longer a mere oscillation, measured mathematically by an epicycle; it had some component outside of the radius to the sun. The epicycle, as a purely mathematical model, did not explain this component. It had to be given a stronger interpretation if the theory were to be taken seriously, so that it would explain what was preventing the planet itself from obeying the distance law.

Kepler did not supply one, but insisted that the epicycle had no physical reality, and that its imagined center moved only on account of the planet's motion. The empirical failure of this theory absolved him from the need to supply a precise description of what was happening to the planet. The model of Chapter 49 remained an abortive attempt at separating the components of his theory. We are left with the question of why he ever undertook such an ill-advised modification of his earlier work. To answer it we must turn our attention, insofar as we can, to the little-understood regions where, for Kepler, the scientific imagination merged with the religious.

Let us examine more closely Kepler's reasons for dissatisfaction with his previous formulation of the epicyclic distance model. He could not compute

---

[105] *G. W.*, 3: 311: 40–312: 1. For Kepler, remember, forces caused motion rather than acceleration. The planet's velocity from its own moving virtue could be simply added to its circumsolar velocity (even though each of these motions was curvilinear), just as a Newtonian physicist would add accelerations.

positions without knowing the perimeter of the oval orbit. This depended on the shape of the oval, which was determined by the simultaneous working of the distance law and the epicyclic model. The oval and its perimeter were not determined by any rule, but only by the cumulative effect of two distinct processes. It was defined physically, but not geometrically. "And this defect is not even of our understanding, but is quite alien to the original arranger of the planetary courses: heretofore we have not found such an ungeometrical conception in his other works."[106] Kepler was completely serious about this argument. His scientific work, profoundly original, and fundamental for the development of modern science, was based on the supposition that the universe had been created according to a rational plan. Kepler's highly individual version of this commonplace presumption, although it colors his whole work, must be sought most particularly in the *Harmonice Mundi*. There it remains largely unknown, even though it has become trite to describe Kepler in general terms as a Platonist. This is not the place to attend the harmonies of Kepler's world, but we must grant that their appeal to him often took precedence over his desire for physical understanding.

So far as we can tell, the dissonance which had troubled him in Chapters 46–48 arose from the oval being defined only incrementally, from the instant-by-instant interaction of the two virtues. Such a curve is entirely satisfactory by modern notions; and Kepler himself had managed to calculate positions on it. But how could it have been created that way? The solar and planetary virtues were of just the right strength to yield a reentrant orbit: the period of circumsolar revolution equaled the period of variation in distance.[107] The former period, however, depended upon the length of the path, which was, as Kepler said, an ungeometrical conception (*anticipatio*: a preexisting concept). The Creator could not have built it into his plan for the universe. The two forces could not have been tuned without guessing the length of the oval, and Kepler had never found creation to be so disorderly as that.

The corrected oval of Chapter 49, despite its shaky physical basis, could readily be determined by divine geometry. The distance law governed a point moving on a circle. The circumference of this point's motion, and hence the mean increment of motion, were geometrically known, so that the motion at any given distance followed from a simple inverse proportion. The theory was conceivably true. After calculating the equations, and finding them worse than his earlier theory, Kepler remarked that he could have rested content with his vicarious hypothesis if he were merely concerned with numerical accuracy. His new astronomy demanded more.

Kepler resumed his struggle with the Martian orbit by gathering together, in Chapter 51, his calculations of the distance of Mars from the sun at pairs

---

[106] *G. W.*, 3: 310: 31–34.

[107] Notice that this resonance really was something to be explained, until Newton succeeded in explaining it. With Kepler's faulty concept of inertia, an eccentric orbit required two distinct forces, and the stability of the apsidal line was adventitious, and not unreasonably attributed to the Creator.

of positions symmetric to the apsidal line. He computed these distances not from theory but from comparison of observations separated by integral revolutions of Mars. (Some of them had been used in Part III when he was investigating the earth's orbit, and had incidentally determined the distance to points on the Martian orbit.) To each position of Mars in the "descending semicircle," from aphelion to perihelion, he paired the corresponding position in the "ascending" semicircle. The distances in each pair were the same within the precision of the calculations.

(Kepler estimated the error of his results, roughly, by comparing different computations of the same distance. He found it to be about 200 parts, where the radius of the *earth's* orbit is 100,000. After confirming the equality at six pairs of points, he concluded in the next chapter that his apsidal line was correctly placed.)

This was not so trivial a conclusion as it may seem. Kepler's apsidal line passed through the sun, rather than the mean sun as required by all previous planetary theory. He had adopted the sun itself as the intersection of all the apsidal lines in Chapter 6, on the basis of sound, if very general, physical considerations. Since then he had used it consistently, and one is a bit surprised to find the matter raised again. All along, though, the use of the true sun had been an innovation. Kepler was quite conscious of it as such, and moreover as an innovation which still lacked proper astronomical justification. Now at last he had provided that justification. As he had shown in Chapter 5, a Martian apsidal line constructed through the mean sun would pass very nearly through the same equant center as the one he had constructed through the true sun. The mean sun was distant from the true sun by 3600 parts, where the radius of the earth's orbit was 100,000. Therefore the orbital centers required by the two apsidal lines, being halfway from the shared equant to the respective suns, would differ by 1800 parts. The whole Martian orbit would thus shift by 1800 parts if the apsidal line were placed through the mean sun. This displacement would be in a direction more than 50° away from Kepler's apsidal line. The symmetry of his computed distances, whose error was perhaps 200 parts, could not withstand so great a disruption. (On Figure 21—not to scale—E is the Martian equant center. Shifting the apsidal line from S, the sun, to $\bar{S}$, the mean sun, would displace the orbital center from C1 to C2, a distance of 1800 parts. This would make it perceptibly asymmetrical about Kepler's apsidal line AP.) Tycho's observations, which now had shown the orbit to be symmetrical about the apsidal line through the true sun, would show it not to be symmetrical about one through the mean sun. Direct computations were unnecessary; the matter was settled.[108]

---

[108] "...It is a disgrace for an astronomer to try out something with numbers whose foundation he has not already seen in geometry...," *G. W.*, 3: 336: 32–33. In his letter to Maestlin dated 5 March, 1605, Kepler referred to Chapter 51 as Chapter 51, indicating that by that date he had assembled that much of the *Astronomia nova*: letter #335, in *G. W.*, 15: 171: 62–65. At the beginning of 1605 Kepler wrote to Longomontanus that he had completed (probably these) 51 chapters: letter #323, in *G. W.*, 15: 141: 255–256. This is one of the few indications we have of a completion date for a portion of the book as he finally published it.

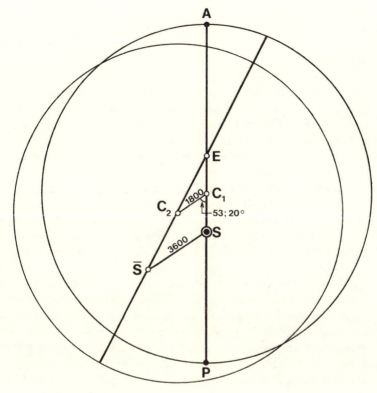

Fig. 21

To improve the precision of his calculated positions of Mars, Kepler developed in Chapter 53 an ingenious construction for use with observations before and after opposition, when that planet was close to the earth. The method was based upon those things in which he was most confident. These were the distances and longitudes of the earth's orbit, little eccentric and hence accurately approximated by his old theory, the short-term increments of Martian longitude, which in the vicarious hypothesis were even more accurate than the longitudes themselves, owing to the absence of accumulated error; and in some parts of the orbit,[109] the increments of distance, which were likewise known more accurately than the distances themselves. From these better-known elements, and observations taken shortly before and after opposition, he could improve his estimates of the distance and longitude of Mars. The distance estimates, as he later revealed, were initially taken from a theory not

---

[109] Near the apsides the distances changed very little in any plausible theory, so that one could not be badly mistaken about the increments. (No comparison of different increments was needed.) Kepler made the curious statement that the increments were also known, reasonably well, in the mean distances. *G. W.*, 3: 337: 33 – 338: 1.

yet described, and much better than his previous one. Indeed, the slight refinements he worked out in Chapter 53 applied mostly to the longitudes of the vicarious hypothesis.[110] Kepler laid out a table of twenty-eight Martian positions for later use.

Let us pause for some remarks on chronology. Chapters 51 and 53 each set out distances and anomalies, refined observational data for the discussions to follow. On closer inspection, however, the two chapters are quite different. Several of the observations used in computing the Chapter 53 positions were among those used in the earlier chapter. The positions, however, had been corrected. There can be little doubt that Chapters 51 and 52 date from a much earlier period than Chapter 53. They were devoted to proving that the apsidal line of Mars passed through the true sun, which was certainly one of Kepler's earliest theoretical results. Of the six values of mean anomaly for which Kepler tabulated the observationally computed distance in Chapter 51, one was specifically employed to establish that the apsidal line passed through the sun itself. Three others were at locations he had used to investigate the earth's orbit in Chapters 26, 27, and 28. In estimating a fifth he remarked that a discrepancy of 127 parts was negligible in a matter "where 1800 or 3600 parts were at issue."[111] This was a sensible point to make when trying to show that Mars' apsidal line passed throught the true sun, for the true sun was distant from the mean sun by 3600 parts, and the corresponding orbit was shifted by 1800 parts, as we have seen. It would not have been a pertinent thing to say during calculations aimed at determining the shape of the Martian oval, where much smaller distances were critical. Finally, as we noted above, a few but by no means all of the Chapter 51 positions were included among the twenty-eight tabulated in Chapter 53; and those few had been further revised.

The positions in Chapter 51 were selected because observations were available at the symmetric position in the other half of the orbit; and they were calculated with the techniques employed in Part III to determine the earth's orbit. Those in Chapter 53 were selected to form clusters of four around each of the oppositions between 1582 and 1595; and they were refined with a subtle technique which presupposed both the theory of the earth, and a later theory (from Chapter 56) of the distance of Mars. It seems clear that Chapters 51 and 52 were inserted out of chronological sequence, and in fact represent an earlier stage of Kepler's work. Chapter 53, in contrast, was finished for publication at a time later than the work it contains.

With all of these positions in hand, Kepler was able to discard his working model for the distances to the oval of Mars. Since Chapter 45, he had assumed

---

[110] W. Donahue has shown that the numbers Kepler published in Chapter 53 were not all derived from any one set of assumptions, and that the table published in that chapter probably represents Kepler's attempt, after discovering his final distance law, to smooth the exposition leading up to it. "The Peccadillo of Johannes Kepler," read at the meetings of the History of Science Society, Bloomington, Indiana, 2 November 1985.

[111] *G. W.*, 3: 329: 2–3.

an oval[112] where at mean anomaly α the distance of Mars was proportional to the square root of $(1 + 2e\cos\alpha + e^2)$. Trivial computations showed that this theory of the distances was incompatible with the positions calculated from Tycho's observations. The epicyclic distance theory pulled the plane too close in the mean distances, where the original circular orbit had left it too far away.

These direct computations of distance explained, within the context of Kepler's physics, the errors he had already found in equations computed from his distance model.[113] He had found that model to move the planet too fast in the mean distances. No wonder: for it brought the planet too close to the sun, into a region where the virtue of the sun's rotating image was stronger. As a distance model, the uniformly rotating epicycle was inadequate. It was appealingly simple, and it yielded an oval orbit. The planet's oval, however, was less constricted. Kepler needed another theory for the distances.

Both lines of reasoning used to disprove the old distance model, the calculated distances and the equations, suggested that the lunula which was cut off between the oval and the circle needed to be about half as wide as that model made it. The greatest width of the lunula had been 858 parts, where the Martian orbital radius was 100,000. The oval orbit fell short of the circle, Kepler had learned, by only about half this much in the mean distances. Moreover, the physical equations computed from the distance law and the oval model erred by about the same amount, but in the opposite direction, as the physical equations from the circular orbit. Kepler needed a distance theory that would pull the planet inside the circle by roughly half as much as before, thus by about 429 parts or thereabouts.

Kepler had been working on lunar theory when he realized that his oval was too constricted. His first inclination was to suppose that the epicycle governing the planet's distance was accelerated in the apsides, in (strained) analogy to the lunar "variation" discovered by Tycho, which moved the moon faster in the syzygies. Already, however, he was thinking about the symmetry which the observations indicated in the oval—as if the oval were an ellipse.[114] Pursuing this lead, rather than the physics of the epicyclic distance theory, he soon found a way to produce just the kind of symmetric oval he needed.

In Chapter 56 he explained how he came upon the key relation in such a theory. Despondent about the failure of his very arduous labors on the oval orbit, he accidentally noticed the value of the secant of 5;18°, the maximum optical equation of Mars. This secant is 100429, where the radius is 100,000. The coincidence of numbers struck him, and he reflected that the lunula, like

---

[112] Actually, he had used two ovals with this distance relation, that of Chapters 46–48 and 50 and that of Chapter 49.

[113] C. Wilson has correctly emphasized both the impossibility of determining the true orbit from distance calculations alone, and the role of the equations of center alongside the rough distance calculations in suggesting the kind of improvements needed: "Kepler's Derivation of the Elliptical Path," *Isis* 59 (Spring 1968): 4–25.

[114] Kepler to Fabricius, letter #308, in *G. W.*, 15: 79: 71–80: 1.

the optical equation, was greatest in the mean distances, and its width there was precisely the excess of the secant over the radius, 429 parts. One could substitute the radius for the secant at the quadrant—and indeed, more generally throughout the orbit—and obtain an oval orbit of about the right dimensions.

This critical step toward the ellipse, which as Kepler tells it is charmingly fortuitous, still leaves quite a lot unexplained. He was describing the circumstances only approximately. On a circle with eccentricity of 9265 parts (a value he had just confirmed) the maximum optical equation is not 5;18° but 5;19°, and it occurs not at the quadrant, where the eccentric anomaly is 90°, but lower, where the true or coequated anomaly is 90°. On any oval it would be still greater. What does equal 5;18° on that circular orbit is the optical equation precisely at the quadrant, where the eccentric anomaly is 90°; and the secant of this angle is indeed 100,429. We make these minute distinctions, not because they are themselves of any importance, but because for Kepler himself, at this time, the numerical coincidence really was just a coincidence. His needs were not that precise; so he gave a rough description, and not a precise one. He needed some theory that would trim the sides of his circle, and remove a lunula which was greatest around the quadrant and had a width there of perhaps 400 to 450 parts. The excess of the secants was greatest around the quadrants (not exactly *at* the quadrant, for the optical equation increased for a while longer), and at the quadrant it was 429. The precise agreement with half the calculated width of the former lunula caught his eye; but the general agreement both in the magnitude and in the mode of variation (with the maximum in the mean distances) captured his attention.

And yet, why should it have done so? Substituting the radius for the secant gave Kepler a "distance" attached neither to the sun nor to the planet on its oval. If we refer back to Figure 16, the optical equation at point E is angle AEB. The secant of this angle is, in more modern conception, the ratio of EA to EB. For Kepler it was the distance EA simply, where EB is the implicit radius of reference; and he proposed to set the planet's distance from the sun equal to this radius EB, measured from the sun at A to the planet at a point inside the circle, somewhere near E. Taking another distance to the circle, say AH, as a secant, Kepler replaced it with a radius which is found to be HR, where R is the foot of a perpendicular dropped from the sun onto the diameter originating at H. Again, this distance is to be measured from the sun at A to some not very well defined point inside the circle, near H. It is a strange way to construct an oval, and not a way likely to occur to someone who has simply been thinking about distances from the sun to the orbit.

It occurred to Kepler because he had used these distances before. When trying to calculate the area of the conchoid that was almost equivalent to the aggregate of distances to a *circular* orbit (Figure 17b), he had constructed a second conchoid (Figure 17c) as one which really was equivalent to the area of a sector. (We discussed these conchoids on pp. 82–87). To achieve this equivalence, we recall, he had been forced to replace the actual distances with their projections on the diameter, because these projections were perpendicular to the circumferential arcs which he was treating as the bases of "Archimedean triangles." He had in fact replaced the secants by their radii, with respect to an angle that was exactly the optical equation for the circular orbit; and this with no purpose beyond estimating the error in his area law, as an approximation to the distance law. Only now did he realize that those shortened distances, laid out along diameters of a circle which was not the orbit, were just about right to extend from the sun to the oval that *was* the orbit. The impact of this realization, as he wrote, was like awakening from sleep.[115]

Kepler did not remark here, although it must have been obvious to him from his earlier work, that the areas of circular sectors with vertex at the sun *exactly* measured the sum of however many of these new distances were contained in them. (In our discussion of Chapter 40 we showed how the second conchoid was constructed for the specific purpose of attaining an exact proportionality.) That convenient equivalence had no bearing on the physical plausibility of his new distance theory, nor upon its empirical adequacy. Indeed, mentioning the fact at this point, with the new theory as yet unjustified, would not so much have supported the theory as cast doubt upon Kepler's reasons for introducing it. Instead he continued the chapter by pointing out the satisfactory way in which this new distance theory agreed with his physical considerations, and with the distances calculated from observation.

For the physical aspects he referred back to the general discussion in Chapter 39, and to the epicyclic diagram which had accompanied it (Figure 15b). Because of the equivalence of that epicycle to an eccentric circle (Figure 15a), $\alpha\delta$ on the epicyclic diagram was the distance to a point on an eccentric circle where the eccentric anomaly equaled angle $\gamma\beta\delta$. By the same equivalence, angle $\delta\alpha\beta$ was the optical equation. Hence to substitute the radius for the secant was simply to take the distance as $\alpha\kappa$ rather than $\alpha\delta$. At other values of the eccentric anomaly the distance was likewise defined by dropping a perpendicular from the epicycle onto its main diameter $\alpha\gamma$. The planet was "librating" in the diameter of the epicycle. This theory pulled the planet inside the circle, since $\alpha\kappa$, the "diametral" distance, is less than $\alpha\delta$ or $\alpha\iota$, the "circumferential" distance. Furthermore, these distances had none of the baffling irregularity of the ones extending to the epicyclic circumference. The increments of distance were the same at equal removes from aphelion and perihelion, as $\gamma\kappa = \mu\zeta$. Although longer in the mean distances than near the apsides, the new increments were well enough behaved to have arisen from some natural, physical process.

Kepler also had to test his nascent theory of diametral distances against the calculations from observation. The distances in Chapter 51 were tabulated for given values of mean anomaly, while his distance theory was now based on eccentric anomaly. He therefore had to subtract the physical equations

---

[115] *G. W.*, 3: 346: 2–4.

before trying out the new model, and he had never been able to pin down the physical equations precisely. No matter, though: an error of eight minutes in the equation altered the distance by no more than a very few parts of a hundred thousand, well within the error of his calculations. At all six pairs of positions from Chapter 51 the agreement was reasonably good. As to the twenty-eight positions tabulated in Chapter 53, Kepler now revealed that the distances he had postulated, and then confirmed, by the method of that chapter had been derived in the first place from this theory. Their confirmation was an accomplished fact.[116]

Now that he had a mathematical theory of distances which agreed with the planet's real motion, as determined from observations, Kepler again turned to the problem of physically describing how the planet went about the task of changing its distance in just that way. (His letter to Maestlin of 5 March, 1605,[117] was evidently written after he had arrived at the distance theory, and after he had worked out his physical theories in outline, but before he had managed to derive the distances from the physics.) The librations of the new theory, greatest in the mean distances and symmetrically smaller toward either apse, varied in a believably natural way with the "fortitude" of the eccentric anomaly. The fortitude, he asserted, was in turn about as the sine.

Kepler was not very clear in explaining his meaning here, but the gist seems to be that changes proportional to the sine of a physically defined angle were not uncommon in nature. To show how easily this type of pattern could appear, he postulated a circular river carrying a boat around its course. A steering oar, adjusted smoothly so as to rotate half a revolution in each trip around the center, would guide the boat in toward the center, then back out. In Figure 22, imagine a river flowing swiftly around the circle in the counterclockwise direction. (Ignore the ripples with which Kepler's illustrator adorned the figure.) At D and E the river, pressing on the oar as if on a sail, forces the boat inward toward the center. At perihelion F, the pressure is even and moves the boat neither inward nor outward. At G and H the pressure forces the boat back outward to C, where the oar has made half a revolution from its original position. As Kepler said, "the oar turns around once in twice the periodic time of the planet." Such a boat, like a planet, travels an eccentric but reentrant path, and its distance from the center changes most rapidly at the middle distances.[118]

This analogy, although it showed how easily one could produce motion generally like that of the planets, was not of any real use. The rotation of the

---

[116] C. Wilson ("Kepler's Derivation," p. 12, n. 32) appears to believe that since the libration theory supplied the distances in the table of Chapter 53, these distances, or at least the twenty of them not calculated explicitly in the text, were not observationally determined. On the contrary, the technique of that chapter was a way of using observations to confirm or refine a postulated distance. Although obtained initially from the libration theory, the distances in Chapter 53 really had been confirmed as (approximately) correct by the observations.

[117] Letter #335, in *G. W.*, 15, 170–176.

[118] *G. W.*, 3: 349: 1–26.

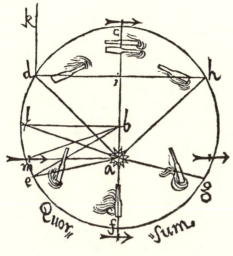

Fig. 22

steering oar, once every two times around the circuit, was fairly simple, but was an unlikely mechanism for the planets, particularly considering that the moon kept the same face always toward the earth. Moreover, the analogy was crude, for an oar used the weight of the water to steer. The river of the sun's image was immaterial; the "oar" should likewise be immaterial.

In fact, Kepler had already argued that planets were carried by "the immaterial image of a magnetic virtue in the sun."[119] Perhaps the oar that regulated the distances was also magnetic, for Gilbert had shown that the earth, one of the planets, was a giant, rotund magnet. Magnets had a faculty of assuming and retaining a direction or orientation, as shown by any compass needle, or indeed by the earth itself.[120] Suppose that a planet was like an enormous magnet whose axis retained a constant direction. Figure 23 shows such a magnet, represented by an arrow, at four points in the planet's orbit. The pole at the point of the arrow was attracted by the sun; the other pole was repelled. At aphelion C the two poles were presented equally to the sun, and the planet, hindered from rotating by its directive faculty, neither approached nor withdrew from the sun. Carried slowly around toward M, the planet increasingly presented its attracted pole to the sun, and began to draw closer. As it reached M, the attracted pole pointed directly at the sun, and the approach was most rapid. Thereafter the approach toward the sun became

---

[119] G. W., 3: 350: 4–5.

[120] Copernicus had hypothesized an annual rotation of the earth to account for its axis being always pointed in the same (tropical) direction. Kepler, whose frame of reference for the earth's motion around the sun was no longer that of a rotating shell, asserted that mere constancy of direction of the axis did not require a special cause, although the very slow precession of the equinoxes perhaps did. G. W., 3: 350: 30–41.

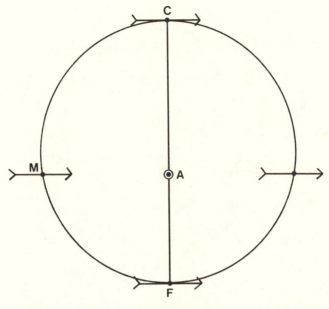

Fig. 23

less pronounced, until at perihelion F the planet again neither moved closer nor away, because neither pole pointed nearer the sun than the other. Exactly the opposite happened in the remaining half of the orbit, as the planet presented its other pole to the sun and was driven off.

The distance theory required that an increment of libration, in an arc near perihelion, be the same as one in an equal arc at the same remove from aphelion, even though the planet would delay longer in the more distant arc near aphelion. Kepler had no mathematical argument to justify his magnetic analogy on this count, but he did know that the attraction of a magnet weakened with distance. It was at least possible, he urged, that the weakened force at greater distance compensated for the longer delay there.

This magnetic model required that the force which retained the poles in their permanent alignment be very strong; but what was to prevent that ? If the tug of the sun on one pole of the magnetic planet, and its push on the other pole, slightly deflected the axis from its pristine direction, the slippage would perhaps have no worse effect than the very slow progression of the apsides which astronomers had in fact observed. Toward aphelion the axis would tend to rotate *in antecedentia*, clockwise in Figure 23, because of the contrary pull and push on its ends. Toward perihelion it would turn the other way: and here the rotation would be faster, because the sun was closer and was therefore pulling harder. The motion of the apsides would be back and forth, but with the forward component predominating in the long run. One cannot help observing, upon reading this ingenious explanation, that Kepler was neglecting what he had himself argued a page earlier, that the planet's longer delays

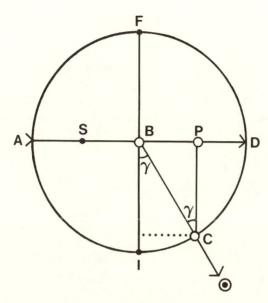

Fig. 24

near aphelion compensated for the weakening of solar influence. The action of the directive faculty holding the planet had not been specified in any detail, however, so we will not criticize too strongly Kepler's attempt to turn a profit on the sun's tugging at the magnetic axis.

In order to develop the mathematics of the magnetic or quasi-magnetic force governing the planet's libration, or motion toward and away from the sun, Kepler needed to postulate the manner in which the libratory force varied as the planet was carried around the sun. Writers on the magnet had shown that its force divided regularly with the division of the magnet itself, so he felt free to consider a single axis in the planetary body. In Figure 24, then, let the circle represent the body of the planet. AD is the axis of the libratory force, D is the pole attracted to the sun, and A the pole repelled by it. If the planet were at aphelion, and the sun in the direction BI, the two poles would make equal angles with the radius from the sun and no libration would ensue. Let the planet move away from aphelion, until the sun lies in the direction BC. Here the true or coequated anomaly is angle PCB or CBI, which we shall denote γ. Now the sun is exposed to the attracted pole at the acute angle CBD, and to the repelled pole at the obtuse angle CBA. The attractive force overcomes the repulsive force.

The net effect, Kepler argued, would follow the proportion of scales: the attractive force would be as AP, the repulsive force as DP, where P is the foot of a perpendicular dropped from C to the axis. For this was the way things worked in nature. If AD were a balance beam suspended from the arm CB (you must rotate the diagram about 140° to visualize this), the arm and beam assume the given angle if the weights at A and D are respectively as AP and DP. Moreover, if the beam were suspended from an arm CP (invert the diagram), the arm remains horizontal if the weights at A and D are respectively as DP and AP. (Incidentally, the first of these assertions was false. The second, generally known as the law of the lever, was correct.)

If one attributed this natural and plausible form to the libratory force, then the net force on the planet was as the difference of AP and DP. Marking off AS equal to DP, this difference was simply SP. Since SP was always twice BP, the net force varied as BP during the planet's revolution about the sun. But BP, or CN, was simply the sine of the coequated anomaly γ. Magnetic planets of the kind Kepler had hypothesized were attracted to the sun, or repelled by it, as the sine of the coequated anomaly, other things being equal.

How, then, to express the cumulative effect of this libratory force? Kepler wrote this physical analysis after thinking of the distance theory presented in the previous chapter, and so he already knew approximately the answer he needed. At aphelion on the epicyclic distance diagram, the planet's distance from the sun was $1 + e$; at eccentric anomaly β the distance was $1 + e \cos \beta$. The libration, or decrease in distance, was thus $e \cdot (1 - \cos \beta)$, or simply $e \operatorname{vers} \beta$. He needed to show, therefore, that the sum of a great number of increments of libration, each proportional to the sine of the coequated anomaly, would grow at the versed sine of the eccentric anomaly. His argument to this effect was more than a little obscure, but can, I think, be made intelligible.[121]

First Kepler tackled the mathematical aspect of the problem; lacking integral calculus, he could only approach it numerically. For every angle consisting of a whole number of degrees in a quadrant, he summed the sines for each degree up to that angle and compared the sum to the versed sine of the angle. That is, he tested whether or not, for some K,

$$\sum_{i=1}^{n} \sin i° = K \cdot \operatorname{vers} n° \tag{24a}$$

He found this proportionality to hold with reasonable accuracy, except at the very beginning of the quadrant where the relative errors were large. (Doing physics without integrals is not easy. The integral formulation of equation (24a) holds exactly; Kepler's one-degree increments were simply too large to give a good approximation at the beginning of the quadrant.) Since the observations had only demonstrated that the libration was approximately as the versed sine, and the absolute amount of error was always small, these calculations were perfectly satisfactory to Kepler. In evaluating his argument here, we must not be led astray by our knowledge that in the final theory the

---

[121] Kepler was for a while concerned over the disagreement between his physics, which indicated a libratory force proportional to the sine of the eccentric anomaly, and his distance law, which indicated that the completed libration was in fact proportional to the versed sine of that angle, until he realized that the latter represented the cumulative effect of the former. *G. W.*, 15: 252–255.

libration was precisely as the versed sine, and that the integral version of equation (24a) is exactly true. Kepler did not know either of these things when writing Chapter 57, and was probably not too distressed at the imperfect agreement of an approximate calculation from physics with a theory known roughly to agree with observation. The difference certainly lay within the limits of observational and computational error.

A more serious problem remained. He wanted to set the sum of sines of the *coequated* anomaly proportional to the versed sine of the *eccentric* anomaly. This consideration seems to invalidate completely the calculations represented by (24a); for the difference between the two anomalies is the optical equation, which depends upon the eccentricity. No amount of mathematical elaboration will make (24a) exact where the angles are different on either side of the equality, and the amounts of difference vary among the several planets. Kepler did not need to adjust the mathematics, however, for the physical situation was more complicated than he had yet taken into account.

He divided this problem into two. In order to see the distinction, let us restate (24a) so that the summation is done (properly) with the coequated anomaly, but the versed sine is still that of the eccentric anomaly. For eccentric anomaly i, let the optical equation be d_i, so that the coequated anomaly is $(i - d_i)$. We can then sum the sines at integral degrees of coequated anomaly:

$$\sum_{(i-d_i)=1}^{n} \sin(i - d_i)^\circ = K \cdot \text{vers } n^\circ \qquad (24b)$$

In forming this cumbersome equation—whose use will shortly become clear— we have altered the left side of (24a) in two ways: the index of summation and the argument of the sine function are both smaller. As Kepler put it,[122] in computing the sums for a quadrant by (24a) instead of (24b), he was summing *more* sines than it seemed he should have, and each of the individual sines was *larger* than it should have been, according to his physical theory. He went on to argue that the first of these apparent errors was physically justified.

The key to this argument is that he had supposed the libratory force to be proportional to the sine of the coequated anomaly if other things were equal. In arbitrary pieces of the orbit, other things were not necessarily equal. He had already remarked that the force of libration weakened, like that of a magnet, at greater distances. Furthermore, any force would move the body farther if it had a longer time in which to act. According to the distance law, the time was proportional to the distance for path elements of any fixed length, so these factors cancelled in a path element of any length. It was only necessary to keep the path elements of the same length, so that the sines representing successive increments of libration would be compared to constant increments of the planet's motion about the sun. Degrees of coequated anomaly did not represent equal path elements, since the element subtending a one-degree arc

---

[122] *G. W.*, 3: 354: 33 – 355: 2.

was obviously longer at a greater distance from the vertex.[123] Degrees of eccentric anomaly, on the other hand, corresponded to equal (or nearly-equal) elements of motion about the sun. Kepler had been right, then, to sum his sines at integral degrees of eccentric anomaly. His physics should have meshed with his distance theory, not according to (24b), but according to

$$\sum_{i=1}^{n} \sin(i - d_i)^{\circ} = K \cdot \text{vers } n^{\circ} \tag{24c}$$

In performing his calculations according to (24a) rather than (24c), therefore, Kepler had erred only in accumulating sines of angles that were too large. He did not think this error very serious.[124] At the beginning of the quadrant the equations d_i were small; and at the end of the quadrant, where the equations were large, the sines differed by little even with an optical equation of several degrees. Moreover, the substitution of coequated anomalies $(i - d_i)^{\circ}$ for the eccentric anomalies i° would tend to lessen the inexactitude he had found in his trial computations with (24a). He had found the partial sums to be slightly too large, relative to the versed sines, so taking the sum of smaller sines would reduce the error.

In fact, Kepler's whole argument that his physics was adequate to account for the success of the versed-sine libration theory was explicitly written in approximate terms. The numerical proportionality (24a) was roughly true, and the use of eccentric instead of coequated anomaly was acceptable, compensating somewhat for the previous discrepancy. He concluded that he had brought the proportionality within the limits of perceptible error (*intra sensus propinquitatem*).[125] The running abstract of his argument in the margin stated that the proportionality held *valde praecise*, but this abstract was written later, when Kepler was confident of his final theory. Kepler did not know, when writing Chapter 57, that in his final theory the versed-sine relation, and its derivation from his physics, would be exact. At that time, we may speculate, he was not certain whether the versed-sine theory was inexact, or his mathematical abilities were deficient, or physical details as yet unknown entered the process.

One point of detail about which we would like to know more is the very abbreviated argument that the sines should be accumulated for integral degrees of eccentric anomaly. This formulation, even more than the versed-sine libration theory itself, appears to have been one that was better than Kepler knew when he arrived at it. Kepler took sines at each degree of eccentric anomaly so that the libration which each sine represented would correspond to a constant amount of motion around the sun. As we have remarked, he had not been certain since abandoning the circular orbit just

---

[123] Aiton apparently missed this consideration, which is left largely implicit in Kepler's very brief argument in G. W., 3: 354: 27–40. See "Kepler's Second Law," p. 83, n. 51, and "Infinitesimals," p. 298.

[124] G. W., 3: 355: 3–6.

[125] G. W., 3: 355: 14.

what was meant by eccentric anomaly. He had not addressed the question, even when proposing that the completed libration grew as the versed sine of the eccentric anomaly. As he said then, several ways of calculating the physical equation were accurate within eight minutes or so, and such errors, although unacceptable in longitude theory, permitted distance calculations well within the error of the distances he could calculate from observation. I think we may presume, from the logic of Kepler's argument here, that by eccentric anomaly he meant in this passage not an angle of any kind, but arc length along the orbit. Yet even here lurks an ambiguity. An arc of the orbit is described by the simultaneous action of the rotating virtue of the sun and of the libratory force. Sines representing increments of libration should be proportioned not to equal arcs of the orbit, but to equal components arising from the rotating image, and thus perpendicular to the libration.[126] Equal increments of eccentric anomaly, as Kepler defined it later, were equivalent in precisely this sense. It is far from certain, however, that he had formulated his new definition of the eccentric anomaly when he wrote this section. If we may trust the ordering of Chapters 57–59 he had not. To understand Chapter 57 it suffices that we recognize the calculations of this and the previous chapter as approximate. The length, whether total or circumsolar[127] only, of a one-degree arc of eccentric anomaly was always just about the same under any definition of the latter. A one-degree arc of coequated anomaly varied in length by over 18 percent, so that sines taken at such intervals would have been badly out of proportion. This was the basis for Kepler's justification of (24c) instead of (24b).

Kepler was generally satisfied with this physical libration theory. It supposed only the working of forces similar to those found naturally in magnets, without requiring any kind of mind to supervise the motion. The theory had one glaring flaw, however. The magnetic axis of the planet had to maintain a constant direction, perpendicular to the apsidal line. In the case of the one planet which could be observed closely enough to bear on this point, such an orientation appeared impossible. Only one axis in the earth's body maintained a constant alignment for even a fraction of a day: the axis of rotation. This axis was directed toward the beginning of Cancer, which was quite close to the earth's apsidal line, and certainly not perpendicular to it. Moreover the equinoxes precessed, so that the apsidal line did not remain in any kind of constant orientation to the axis of rotation. This difficulty was perhaps not

---

[126] Aiton believes that, at this time, Kepler conceived the sun's rotating virtue to act along the oval path, "as if the magnetic force just steered [the planet] in the right direction" ("Infinitesimals," pp. 298–299). Certainly Kepler realized later that the circumsolar force should be responsible only for the component of motion perpendicular to the radius. I believe that he had not worked out this distinction when he wrote Chapter 57, and that we should regard his later revision as mathematical rather than physical; but Aiton may be correct.

[127] I use the word "circumsolar" to describe the action of the sun's rotating image, which urged the planet *around* the sun, in a direction perpendicular to the radial libration. This has not quite the same sense as "tangential," since the orbit is eccentric.

insoluble, but obviously did make the magnetic model less plausible than it otherwise would have been.

Kepler was not entirely committed to the new theory. The magnetic hypothesis was particularly simple: the changing circumstances of an axis whose direction remained constant had yielded the correct variations in distance. If the axis could not hold its direction, he was willing to consider a more elaborate theory, wherein a mind of some sort controlled the planet's libratory motion. With the aid of a faculty either natural or animal, such a planetary mind could perhaps govern the change in distance in some manner about which Kepler did not speculate. He did consider the more general physical question of how a mind in the planet could obtain the information needed to manage properly its motive faculty. The pattern by which the planet was guided, after all, was what he needed to build his true astronomy on physical principles.

The only apparent way for a mind in the planet to perceive its distance from the sun was by examination of the solar diameter. Kepler needed a plausible way to connect this sort of observation with his distance theory, which put the completed libration proportional to the versed sine of the eccentric anomaly. In fact, he found quite a simple connection.[128] As the planet moved from aphelion to perihelion, the change in the apparent diameter of the sun, as seen from the planet, was proportional to the change in the versed sine of the coequated anomaly. This was a remarkable result. According to the distance theory, if the radius of the circle circumscribing the orbit was 1 and the eccentricity was e, then the planet's distance from the sun was given by

$$R = (1 + e) - e \cdot \text{vers} \, \beta$$

$$R = 1 + e \cdot \cos \beta \qquad (25)$$

where β is the eccentric anomaly. Kepler now asserted that changes in the apparent diameter (which varied as $1/R$) were proportional to changes in vers γ, where γ was the coequated or true anomaly.

He did not attempt to prove the assertion in general. Instead he merely showed that it was true at 90° of coequated anomaly, that is, that the size of the sun, as seen from the planet when the latter was 90° from the apsidal line, was halfway between the size seen at aphelion and perihelion. Once again, though, Kepler's proposition was uncannily accurate: on an elliptical orbit changes in the apparent size of the sun (taken to vary inversely as R) are precisely proportional to the corresponding changes in the versed sine of the true anomaly.[129] I know of no elementary demonstration of this relation

---

[128] This second, "mental" theory of libration in Chapter 57 has gone unremarked in the literature. Perhaps historians have been embarrassed that a scientist of Kepler's stature wrote of planetary minds.

[129] Where β is the eccentric anomaly and γ the true or coequated anomaly, one can easily show that the derivatives, with respect to β, of $1/R$ and vers γ are both proportional to $\sin \beta / R^2$.

which does not involve calculus. Kepler's particular demonstration was based upon Euclid, VI, 3, and cannot be generalized. Furthermore, we have no reason to believe that Kepler knew the orbit to be elliptical when he wrote this section of Chapter 57. For the orbit first described in Chapter 58 (the *via buccosa*) the relation is inexact. In fact, the versed-sine libration theory, or its equivalent distance hypothesis (25), did not have any precise meaning so long as Kepler had not decided upon the geometrical significance of the eccentric anomaly β. As we have argued, Kepler was probably unclear about the eccentric anomaly, and he had no motivation to insist that the libration was exactly as its versed sine, when he wrote Chapter 57. He may have simply noticed that both the apparent solar diameter and the versed sine of the true anomaly γ reached their midpoints when the planet was a little inside its mean distance, and then calculated that the midpoints coincided at $γ = 90°$. The precise differential result mentioned in the last footnote may be as accidental to this chapter as the precise integral result corresponding to equation (24a). In the years of his conquest of Mars Kepler's work was fruitful to a degree beyond our ability to explain, and these are but lesser examples of his bewildering harvest.

Whatever the basis of this second versed-sine relation, and regardless of whether Kepler thought it to be exact, it was just the kind of pattern he looked for in the physical world. It permitted him to sketch out an alternate theory of the changing distance, one in which the motion to and from the sun was regulated by a mind rather than by the blind working of natural forces. The amount of libration was itself not amenable to such a theory, for it was tied to the eccentric anomaly. This was not an angle which a planetary mind could perceive, he pointed out, for the center of the orbit was a vacant point.[130] No planetary mind could constrain the libratory increments to vary as $\sin β$, nor ensure that the completed libration was as vers β, for the eccentric anomaly β was not given physically at the location of the planet. The empty center could play no part in determining the planet's motion. On the contrary, it was only by the motion that the center was defined.

The coequated anomaly γ, on the other hand, was defined by the planet, the sun, and the apsidal line. In principle Kepler saw no reason why a planetary mind could not perceive stars marking the position of the apsidal line, for stars were real bodies like the planet and the sun.[131] If such a mind existed and could perceive this angle, or more precisely the versed sine of this angle, it would have all the information needed to regulate properly the planet's distance from the sun. As the planet was carried around the sun, the versed sine of the coequated anomaly would change. The mind would merely have to make sure that the planet moved toward or from the sun so that the apparent diameter of the sun changed proportionally to the change in the versed sine.

---

[130] At this point Kepler used the eccentric anomaly as it had been defined on a circle, as the angle between aphelion and the planet, measured at the center of the circle. *G. W.*, 3: 358: 38–40.

[131] *G. W.*, 3: 359: 3–7.

This theory was not at all like the one using magnetic forces which he had advanced a few pages earlier. Here the actual distance traveled by the planet in its libration did not matter at all: the planet simply traveled until the solar disk looked right. The eccentric anomaly β, so hard to pin down on an oval, was irrelevant, and there was no need for the planet to maintain a constant orientation of any magnetic axis. Corresponding to these advantages there were weaknesses in the new theory. If one no longer had to deal with the distance traveled by the planet, and its unnatural dependence upon β, that was just because there were no longer any specific means proposed to move the planet. This was a theory of the control mechanism, not of the forces. And if the planet did not need to hold a constant orientation, its mind did have to determine somehow the versed sine of the perceived angle. Rather than endow a planetary mind with the capacity for mathematical calculations, Kepler supposed that it would employ a natural means. As he had argued before, the "fortitude" of an angle was naturally measured by its sine. If there were in the planet extended structures (*tractus*) of some kind which were naturally attracted to the direction toward the sun, but were held in some other direction by an animal faculty in the planet, then there would be a continual struggle between the animal faculty and the natural attraction. By hypothesis the *anima* would overcome the natural attraction. The force it would have to overcome would vary, naturally, as the sine of the angle between the two directions. The planetary mind would know the sine of the angle from the effort exerted by its animal faculty, and could proceed to regulate the apparent solar diameter to that measure, by moving toward or from the sun. (Notice the distinction: the magnetic axis had been held in its orientation by a directive faculty, and as it was attracted or repelled by the sun it carried the planet with it. These *tractus* were held in position by a soul serving the planetary mind. They strove to rotate into alignment with the radial direction to the sun, but accomplished nothing more than indicating, by the strength with which they tried to turn, how far out of alignment they were.)[132]

Kepler thus proposed two quite different theories in Chapter 57 to account for the planet's librations. One relied upon natural forces, magnetic or quasi-magnetic, while the other supposed some kind of primitive planetary mind. The former involved mathematical difficulties in cumulating the effect of the forces; the latter avoided these difficulties by showing how a mind, properly equipped for sensation and control, could determine when it had moved the planet far enough. The magnetic theory required an axis of the planet to point constantly in the same direction; but Kepler had burdened the mental theory with a similar requirement, the *tractus*, in order to furnish an indicator of the sine of the anomaly. What was there to choose? In the first place, as Kepler remarked, the natural theory stood by itself, while the mental theory invoked

---

[132] *G. W.*, 3: 360: 8–32. Since the versed sine of the coequated anomaly was needed, we may presume that the *anima* would hold these fibers perpendicular to the apsidal line. The sine of the angle between the solar radius and the fibers would then be the cosine of the coequated anomaly; and changes in the cosine mirror the changes in the versed sine.

natural forces both to actually move the planet and to determine, in some way such as that just described, how fast the versed sine of the true anomaly was changing. A mind was perhaps superfluous. (Besides, Kepler added, his reader might be less willing than he was to believe in a planetary mind!)

Another difficulty with the mental theory of libration was "a certain geometrical incertitude, which I suspect might be repudiated by God himself, who up to now is found to have always proceeded in a demonstrative way."[133] In a different guise this problem had previously appeared in Chapter 49 (see p. 103), where Kepler had tried to avoid applying the distance law on a circumference whose length was not geometrically defined. It reappeared in the mental theory of libration because of the need for an unspecified natural force which the mind could use actually to transport the planet. This force, Kepler assumed, would act more strongly when the planet was near the sun, like the (alternative) magnetic libratory force, or the sun's rotating virtue. The mind, therefore, would have to take into account not only the changes in vers γ and in the apparent solar diameter, but also the changes in the effect of its virtue at different distances. The approach of the planet to the sun was one of the factors determining how fast the planet should approach the sun. There is nothing inherently wrong with this—it is quite a typical situation in classical physics—but as we noted about Chapter 49, it is not the way Kepler's universe had been created. Only by trial and error could the two forces have been tempered so that they completed their periodic cycles in the same time. It was possible, he observed in concluding the chapter, that they in fact did not complete their cycles in the same time, so that the motion of the apsides ensued from the planetary mind's inability to foresee the exact consequences of its actions. In general, though, it seems Kepler was learning that planetary minds, if we may paraphrase Laplace, were a hypothesis of which he had no need.

Kepler now had a distance law agreeable both to the observationally-computed distances and to his physical speculations. He retained it thereafter as a permanent part of his theory; yet from it he was not immediately able to conclude that the orbit was an ellipse. We have dwelt from time to time upon the reason for this uncertainty: Kepler was not sure, and had not been sure since abandoning the circle, what he meant by the eccentric anomaly. He seemed still attached to two of the interpretations which had applied in the old astronomy. In physics generally, as when calculating with the distance law of delays, or the magnetic theory of libration, he used eccentric anomaly as the distance traveled by the planet on its orbit. In geometrical constructions such as that used in his distance theory, and in the more abstract physics where he considered the information available to a mind, eccentric anomaly was still an angle measured at the center of the orbit, from aphelion to the planet. Both uses were essential to his work, and he had not been able to resolve the distinction between them. Perhaps he had not tried. As we have seen, he was not concerned about the exact way of computing the physical equation when testing his new distance theory, for the observational un-

---

[133] *G. W.*, 3: 362: 14–16.

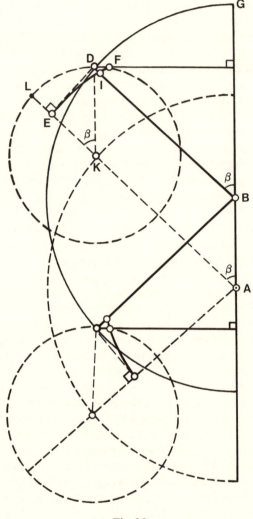

Fig. 25

certainty in the distances exceeded any possible effect of a few minutes' error in the eccentric anomaly—and the difference between the two uses was quite small. Indeed, what could he have done? The orbit was so close to circular that a careful distinction between these two eccentric anomalies would have brought no benefit. As it turned out, neither was the one he wanted.

Kepler recounted his discovery of this last piece in the puzzle in Chapter 58, which was manifestly written after he had the answers.[134] Superimposing

[134] The final stages in Kepler's work on Mars are recounted in a long letter to Fabricius (#358 in *G. W.*, 15: 240–280) dated 11 October, 1605, which reveals little more than the account published in the *Astronomia nova*.

the distance model upon an auxiliary circle which circumscribed the true oval (Figure 25), he constructed planetary positions by treating the eccentric anomaly as the central angle. In the upper quadrant, let the eccentric anomaly β be angle GBD. To determine the distance, we superimpose the epicyclic distance model, shown dashed in the figure. By the basic equivalence theorem for epicycles and eccentrics, β also appears in the epicycle as angle LKD. The libration at β is LE, where E is the foot of a perpendicular dropped from D. (LE is obviously proportional to the versed sine of β in the epicycle. Since the greatest travel of point E equals the diameter of the epicycle, which is also the greatest libration, it is evident that LE equals the libration for this value of β.) LE, then, is the completed libration. From its aphelial distance AG = AL, the planet has approached the sun by the amount LE. It is therefore distant from the sun by AE. Since GBD is the eccentric anomaly, Kepler placed the planet at a point I on line BD, such that AI = AE, the proper distance.

This placement of the planet has been called arbitrary and unjustified,[135] probably on the basis of Kepler's highly self-critical remarks in Chapter 58. In fact, it was the natural and obvious position. His new distance theory told him how far the planet was from the sun, for given eccentric anomaly. And his construction was the most straightforward imaginable: he marked off the distance (AE or AI) along the line indicating the eccentric anomaly (BD). Whatever anxiety he had about the precise use of the eccentric anomaly, he certainly had no reason to expect the radical redefinition that was required.

Kepler devoted one moderately long paragraph to explaining the construction of this model—and then in the next paragraph he summarily rejected it. However natural and obvious it seemed, the theory moved the planet too slowly near aphelion, and too swiftly near perihelion, as Kepler discovered in calculating the physical equations, both with the distance law and with its near-equivalent, the area law. He easily showed that the orbit itself was broader near aphelion than near perihelion (therefore he called the path *buccosa*, cheeky). These errors had the physical implication that the orbit should be more symmetrical, relatively narrower near aphelion and relatively wider near perihelion.[136]

After revealing these facts, Kepler briefly explained how he corrected the orbit. The story is well-known and engaging, but leaves us still in some perplexity. Abandoning the physical libration theory, he thought back to his

---

[135] For example, Koyré, *Astronomical Revolution*, p. 259 ("a mistake"), p. 261 ("unfounded"); Aiton, "Second Law," p. 83 ("gratuitous"). Aiton later corrected his opinion, recognizing that the placement was "natural": "Infinitesimals," p. 299.

[136] D. Whiteside has calculated that the *via buccosa* was observationally indistinguishable from the final ellipse. At any given time the planet's distance from the sun differs on the two orbits, but because the motion around the sun depends on the distance, the two positions lie within three quarters of a minute of arc in true anomaly. "Keplerian Planetary Eggs," *Journal for the History of Astronomy* 5 (1974): p. 14, and "Newton's Early Thoughts on Planetary Motion," *British Journal for the History of Science* 2 (1964): p. 129 n. 42. Kepler himself never paused to make such calculations. Once he realized how to combine his physics with his distances—as discussed in the following pages—he *knew* that the combination was correct.

immensely laborious efforts to calculate positions from his previous distance theory, the uniformly rotating epicycle hypothesized in Chapter 45. His best calculations from that theory had erred, he recalled, by about the same amount as the circular orbit he had used in Chapter 43, but in the opposite direction. The answer was obvious: the true orbit must lie in the middle of those two hypotheses. This line of reasoning had recently set him on the path of the libration theory in Chapter 56, where he was thinking of the maximum width of the lunula by which the orbit differed from a circle. In the end it had led him awry, to an orbit that was asymmetrical, and whose physical equations erred very much as if the asymmetry were the problem. In reconsidering his results, Kepler tried to avoid repeating this mistake by evaluating the overall shape of the orbit, and not merely the greatest width of the lunula.

The earlier model of Chapter 45 had also been asymmetrical. For his calculations, however—the calculations which were just as far from the truth as was the circle—he had approximated it with a symmetric curve, an ellipse. What he needed now was a symmetric oval halfway between that ellipse and the circle which circumscribed it. Such an oval could only be another ellipse, of course.[137] This is very nice, and we have no reason to doubt that it summarizes Kepler's train of thought at some point in his work. What the story lacks is any explanation of how he realized that an elliptical orbit was physically satisfactory. There were innumerable ovals, symmetric or otherwise, which Kepler could not have disproved observationally.[138] The clinching argument in his adoption of the ellipse was the success he finally had in accomodating it to his physical theories; and that part of the account he did not put in the book. We must recreate it ourselves, if we are to understand how Kepler decided that the real orbit was an ellipse.

First let us see the solution itself. The key was a simple but arbitrary reinterpretation of the eccentric anomaly. The new libration theory implied that the planetary distance was AE, in Figure 25, for eccentric anomaly β, where β is angle LKD on the imagined epicycle. The location of the planet itself, however, depends upon the significance one attached to β as a descriptive parameter of the orbit. The *via buccosa* resulted from the obvious interpretation of β as the central angle GBD. It was wrong. If, however, the planet were at F on the perpendicular from D to the apsidal line, when its distance was AE, then the equations were satisfactory. The orbit implied by this construction was an ellipse, as Kepler went on to show in Chapter 59, after the puzzle was solved. His breakthrough, then, was in identifying the eccentric anomaly in the libration theory with the central angle, not to the planet, but to the projection D of the planet perpendicularly from the apsidal

---

[137] C. Wilson has quite properly emphasized that Kepler's distance calculations did not nearly suffice to define the orbit, and that an essential element was his realization that the circle and the earlier ellipse gave opposite and roughly equal equal errors in the equations of center: "Kepler's Derivation," *passim*. I cannot agree, however, with his implication that this realization sufficiently determined the shape of the orbit (p. 18).

[138] See the discussion in Whiteside, "Keplerian Planetary Eggs," especially pp. 13–15.

line to the circumscribing circle. This angle is not of any obvious interest, which makes Kepler's use of it so remarkable. To repeat: if β is the eccentric anomaly, construct a circle circumscribing the orbit, and hence tangent to it at aphelion and perihelion. The angle β, measured from aphelion, defines a point on that circle. From this point drop a perpendicular to the apsidal line, and the planet will lie on that perpendicular, at a distance from the sun determined by Kepler's libration theory.

Only an unthinking reverence for Kepler's achievement has made such a redefinition plausible to historians, in the absence of supporting arguments. (Such arguments exist, but Kepler did not present them until much later, in his *Epitome*—see pp. 163–165.) Here in the *Astronomia nova* the new interpretation of β was merely shown to satisfy the constraints of observation, in that it yielded a symmetric oval (an ellipse) which lay about halfway between the circle and that other ellipse used in approximating the earlier theory.

Kepler also demonstrated the remarkable and interesting fact that on this ellipse the area law worked *exactly* as a substitute for the distance law, where distances were taken at equal intervals of the new eccentric anomaly. He did not show why anyone would want to take them so. Indeed, the physical role of eccentric anomaly in the fundamental distance law, governing motion around the sun, is entirely absent in Chapters 58–60, and hence remains unclarified in the book as a whole. The new relation between the planet and the eccentric anomaly, which was the final step in Kepler's adoption of the elliptical orbit, permitted him to reconcile his physics with the results calculated from observation—so long as he left alone its implications for his distance law. That relation was, however, neither a deduction from the physics nor a result derived from observation.

We do not point this out to criticize Kepler for failing to write a book different from this one. The *Astronomia nova* did not aim at physical theory or observational adequacy alone, but at the true design of the world, and one cannot deny that it was remarkably successful. The means Kepler employed were often roundabout, however, and particularly so in this final step. He informed his reader only that he "began to recall the ellipses" used in earlier chapters, when he realized how the puzzle fit together. We must return, then, to those chapters if we are to follow his thought. Our reconstruction will be tentative, particularly as the dating of those chapters' composition is an open question.

Let us begin with the chief question. When Kepler put aside the libration theory and the *via buccosa*, and began to *revocare ellipses*,[139] how did he come upon the idea for his redefinition of the eccentric anomaly? What line of thought led him to make an oval by moving the planet inward from the circle perpendicularly toward the apsidal line? We are not speaking of physics here; there was no force moving the planet toward the apsidal line. We are concerned with the geometrical interpretation of the eccentric anomaly. On a

---

[139] *G. W.*, 3: 365: 38.

circular orbit this had been unambiguous: in Figure 25, eccentric anomaly β
pertains to point D. The central angle separating D from aphelion is β; the
arc separating D from aphelion is β. How did Kepler realize he had to abandon
both of these interpretations for his oval, by moving the point corresponding
to β inward along a perpendicular DF to the apsidal line?

This, I think, is the principal step which is omitted from Kepler's account.
We are able to reconstruct most of it with some assurance. Consider the
situation facing Kepler with the empirical failure of the "cheeky" path to which
his new distance theory had led him. He knew that he wanted an oval which
was roughly symmetrical, not only about the apsidal line but between its
upper and lower halves as well. He knew that the departure from circularity
should be about half that of the oval he had developed from the discarded
distance theory of Chapter 45. As he remarked in Chapter 58, the figure
halfway between a circle and an ellipse (for he had approximated that oval
with an ellipse) could only be another ellipse. Something else had bothered
him, though, for this rather pat solution had been available since Chapter 47,
when he laid out his table showing the opposite and roughly equal errors of
those two models. Instead of leaping then to the conclusion that the intermediate orbit must be another ellipse, Kepler had continued to develop his
observational data and his physics. He knew that an intermediate ellipse
would be about right, but he wanted the true theory, expressing the physics
of the heavens and the architecture of Creation.

Nevertheless, when his physics had apparently failed him, he took up
ellipses again. That is, he tried to calculate positions, as best he could, on the
still-unknown oval. For this purpose he had little choice but to approximate
the oval, as he had that first oval, with an ellipse. In order to see why this was
a useful thing to do, we must return to Chapter 47, and consider in a little
more detail the technique Kepler had developed there for calculating positions
on an oval that was approximately elliptical in shape. Much of the technique,
not surprisingly, was adopted for his final theory, and will be familiar to those
who are acquainted with Kepler's discoveries. He had originally developed
the method, however, to use on an ellipse approximating his first oval. In that
context, we find that it has characteristic features which illuminate our present
problem.

The technique was based on a property of ellipses: if an ellipse is inscribed
in a circle, all of the perpendiculars to their common diameter are divided in
the same ratio by the two circumferences. In Figure 26, the ellipse AEFB is
inscribed in the circle ACDB. Construct perpendiculars to the major axis AB
of the ellipse; then EG/CG equals FH/DH, and so on. This relation permitted
Kepler to use elliptical sectors as conveniently as circular sectors. The elliptical
sector AES was composed of the slice AEG and the triangle ESG; and the
circular sector ACS was likewise composed of slice ACG and triangle CSG.
The ratios of the respective components were obviously equal to the ratio of
the altitudes, EG to CG, and hence the ratio of the sectors themselves was
that of EG to CG. But this ratio was constant, regardless of the position of
the perpendicular determining points C and E.

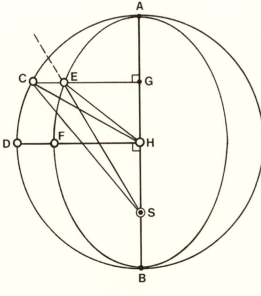

Fig. 26

The constant ratio of circular to elliptical sectors had made it relatively simple to apply the area law to an ellipse, and by approximation to the oval Kepler had been analyzing in Chapter 47. The first step, he wrote there, was to compute the distance, according to the epicyclic model or its equivalent. Next, one ignored the oval shape of the orbit and calculated an area that was proportioned to the whole circle as the mean anomaly was to the whole periodic time. According to the area-law approximation, this would have equaled the area of the circular sector to the planet, with vertex at the sun, if the orbit had been circular. Still supposing a circular sector, one could determine[140] how great an angle (ASC on Figure 26) subtended a sector with the required area. From this angle the perpendicular CG, to the circle, was known. The constant proportionality between ellipse and circle then determined the short segment CE, the width of the lunula at that point, and with this segment one could correct angle ASC to ASE, the coequated anomaly. The constant ratio of elliptical to circular sectors now ensured that the elliptical sector ASE was properly proportional to the mean anomaly. The corrected angle, with the previously determined distance, defined the place of the planet.

Note that in this construction Kepler had not assumed the planet to be at point E, except for the purposes of the longitude calculation. Once that was done, and he knew the planet to lie on line SE, he used his distance theory to determine the final position. (Otherwise his initial instruction, to compute the distance from the epicyclic model, would have been pointless: he did not need

---

[140] This calculation was indirect, requiring the *regula falsi*, as Kepler noted.

the distance for the area-law calculation.) In this earlier theory, in fact, the correct distance did not coincide with the point E on the ellipse which he had used to estimate the anomaly. That auxiliary ellipse had been used only where it was necessary, in the calculation of areas.

Three points regarding this method seem relevant to Kepler's later reappraisal of it, after he had developed the new theory of "diametral" distances. First, the demonstration moved from the circle to the oval by passing perpendicularly inward toward the major axis. Second, the oval was an ellipse, and as an ellipse it was characterized by the property of having a width in constant proportion to the width of the circle. This property was precisely the one which had made it obvious that halfway between the circle and this ellipse lay only another ellipse. Finally, the demonstration here did not address the question of where the distance theory would place the planet, along the line determined by the area law.

Let us, then, appraise these earlier investigations as Kepler would have seen them, upon rejecting the *via buccosa* of Chapter 58 and despairing of his physical libration theory. He knew that the planet moved inside the circle, on an oval path of some kind. Both the equations and the distances he had computed indicated, not too precisely, that this oval lay about halfway between the oval derived from his previous distance theory and the circumscribing circle. If he wished to calculate positions on such an orbit, using the method expounded in Chapter 47, he would again have necessarily used an ellipse to approximate the new oval, for it was only by using the constant ratio of elliptical sectors to circular sectors that he was able to deal with oval areas at all. After using the area law on this new auxiliary ellipse, to determine the coequated anomaly, he would still have needed to place the planet at the proper distance. His best distance theory was still that of the diametral distances, from Chapter 56; and if he used that theory to lay out the planet's distance along SE (Figure 26), he would have found that this time the correct distance reached exactly to point E—the point on the auxiliary ellipse used in the area-law computation.

One can imagine Kepler making this discovery numerically, by comparing diametral distances with those to point E; or more likely, I think, in the course of deriving geometrically the distance to point E. What is surely significant is that Kepler did demonstrate the equality immediately upon resuming his writing, in Chapter 59. It was the first major result in that chapter. The eleventh of the "protheoremata" composing Chapter 59 was introduced with the remark that, with preceding matters established, Kepler could proceed with his demonstration. This theorem showed that the distance from the sun to a point on the ellipse equaled the diametral distance, in the sense of Chapter 56, from the sun to the point on the circle perpendicularly out from the point on the ellipse.

Still another curiosity rests among the computations of Chapter 47. Before applying the area law in the manner just described, Kepler had to estimate the ratio of his (incorrect) ellipse to the circle. He did this in a lengthy construction which yielded, approximately, the width of the "lunula" between the two curves. At the end of the calculation, a marginal note, evidently added later, stated laconically that "this demonstration has its use also in the true physical hypothesis."[141] One must suspect some connection with Kepler's later reworking of this material, in search of an oval halfway between this one and the circle. There does not appear to be enough evidence in the text, however, to establish the connection. An auxiliary point in the construction does lie in the middle of the lunula, and one can show that its solar distance is about that given by the "diametral" distance theory, equation (25), but I am unable to make any plausible argument linking this construction and the discovery announced in Chapter 58.

In sum, then, the digression reported by Kepler in Chapter 58 probably went something like this. Unsure of the exact geometry of the Martian orbit, he returned to the approximative methods he had used in Chapter 47. In order to apply the area law to the oval he temporarily had to assume the oval to be an ellipse. By dividing the ellipse and the circle with perpendiculars from the apsidal line, he was able to compute elliptical sectors as a constant fraction of the corresponding circular sectors. The point of this, of course, was to construct elliptical sectors (as one could indirectly construct circular sectors) with a given ratio to the total area. While locating the planet along the line thus roughly determined, he found it to lie precisely on the auxiliary ellipse he had been using. He had wanted a symmetric oval: now, guided by the perpendiculars he had constructed to measure his sectors, he had found how to obtain one which exactly satisfied his best distance theory, although with a newly reinterpreted "eccentric anomaly." The elegance of the solution recommended it to Kepler's scientific aesthetic; its accuracy, implied by all his earlier computations, decided the question. The long search was over.

Kepler closed Chapter 58 by pointing out, as we have already mentioned, that only an ellipse could lie symmetrically between the circle and his earlier auxiliary ellipse, and berating himself for not having noticed the fact earlier. He devoted the two final chapters of Part IV to some geometrical demonstrations which made explicit the construction of the orbit, and the application of the physical "distance law" by means of the areas of its sectors. These chapters were formal in tone: Kepler was no longer seeking the theory of Mars, but exposing the mathematics of a known orbit.

He began in Chapter 59 by setting out the constant proportionality of the ordinates and sectors of an ellipse to those of its circumscribing circle. He had already done so in Chapter 47, but not very prominently. Then he constructed the point we call a focus of the ellipse, and showed that the distance from the focus to a point on the ellipse equaled the "diametral" distance defined by the focus and the corresponding point on the circle (where the correspondence was determined by perpendiculars from the major axis, or apsidal line). These *protheoremata* connected Kepler's distance theory with the elliptical orbit.

Interspersed with them were other protheoremata, which we may loosely divide into two groups. One group established the area law. Sectors of the

---

[141] *G. W.*, 3: 299, marginal note.

ellipse were proportional to corresponding sectors of the circle; and these were proportional to the sum of distances to (unequal) divisions of the ellipse, provided that the latter corresponded to equal divisions of the circle. In essence this was a restatement of Kepler's early analysis of the area law, in Chapter 40, but now extended to include distances and sectors of the ellipse, in addition to those of the circle. The unanswered question, as we noted on p. 125, was why one would want to divide the orbit unequally in this way. A final group of *protheoremata* wrestled with this question, for which Kepler had as yet no satisfactory answer. He could only show, in general terms, that this particular division of the ellipse was necessary if the distance law was to be equivalent to the area law. The physical question—why one should accumulate distances more closely spaced near the apsides, where the ellipse was more finely divided—lurked behind all this, but remained unasked.

Kepler closed Part Four, and with it his longitude theory, by demonstrating in Chapter 60 computational methods for converting among the mean, eccentric, and coequated anomalies.

The Physical Synthesis

In the fifth and final part of the *Astronomia nova*, Kepler again took up the latitudes of Mars, essentially for two reasons. First, the perfection of his longitude theory enabled him to get better parameters for the latitude theory. Second, he wished to present a physical hypothesis for motion in latitude, and some truly interesting consequences of it. The physical theory itself, proposed in Chapter 63, was uncomplicated, befitting the simplicity he had found in the planets' motion in latitude. The consequences, which he developed in Chapter 68, have received not the slightest notice in the elapsed centuries.[142] They depended largely upon the physical theories of libration and of latitude, both of which turned out to be wrong; but they are nonetheless of great intrinsic interest.

The physical hypothesis of Chapter 63 was analogous to, and perhaps inspired by, the libration theory which governed changes in the planet's distance from the sun. Kepler postulated a magnetic axis[143] in the body of the planet. This remained always directed toward those fixed stars under which lay the limits of the planet's latitudinal motion. The axis or diameter, parallel to itself at all parts of the orbit, steered the planet north and south of the impelling solar image, like a magnet, or like a steering oar on a boat. Its virtue was something like the "directive" faculty attributed to magnets, just

---

[142] Part five of the *Astronomia nova* is more or less neglected in all accounts. Indeed most of it, simply by virtue of its subject matter, falls far short of the startling originality of the first sixty chapters. Chapter 68, however, is very interesting, and I have found no reference to it anywhere.
[143] He refrained from speculating about the structural basis for this "axis" or "diameter." It was not a concrete model so much as an abstract representation of motion in latitude.

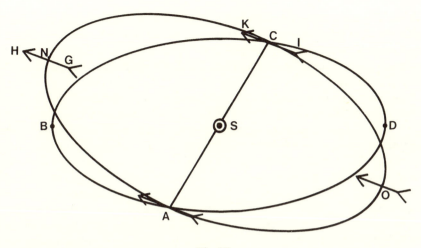

Fig. 27

as the virtuous axis of libration resembled a magnet's "attractive" faculty. As the planet passed its ascending node, point C in Figure 27, its path was tangent to the latitudinal axis, which is represented by IK. The virtue of that axis deflected the planet north of the path of the solar image, the circle CBAD. When the planet reached its limit of northern latitude at N, its path was at right angles to the unchanging direction of the axis, which consequently had no effect on its motion when it was at the limits. At the limits the planet followed the direction of the moving solar image, moving parallel to the sun's equator. As it was carried around toward the descending node A, the south-directed end of the axis took the lead, guiding the planet black to midstream in the river of the sun's image, and on to the southern limit.

Difficult questions remained. Kepler, as always, was improvising his physical explanations. If an inclined axis of some sort guided the planet's digressions north and south of the solar equator, the question remained whether this axis was a natural, corporeal, phenomenon, or whether some kind of mental direction was implicated. These were the same alternatives he had posed in Chapter 57 regarding the axis regulating the planet's librations. Here the case seemed, in general, more decisively in favor of a purely natural explanation. The libratory axis had been attracted and repelled by the sun, and hence had been subject to a force tending to deflect it from its proper direction, as the planet circled the sun. The axis of latitude, on the other hand, steered the planet rather than moving it, and remained always in the same alignment, with no reason to be deflected.

The only peculiarity about this axis of latitude was that when the planet was at the limits of latitude, its axis pointed precisely at the sun: that, in other words, the plane of the orbit passed through the center of the sun. It was by no means necessary in this physics that the sun should be in the plane of the orbit. This is a subtle point, analogous to the physical coincidence which made the eccentric orbit of the planet reentrant. Kepler saw no need to attribute this coincidence to any ongoing mental supervision. The planet had been created with its axis properly aligned, and thereafter had remained, undisturbed, in this alignment through the ages.

By this time, in fact, Kepler was decidedly less enthusiastic than he had once been about planetary minds. Years of physical speculation had brought an increasingly confident belief that natural means alone could explain the patterns of motion for which he had, hesitantly, once invoked mental control. This change, I think, was less a matter of his success in devising physical theories than of a gradual clarification of the problems involved in physical explanation. Kepler was unwilling to propose planetary minds capable of higher functions, such as mathematics. The simple minds he would consider were capable only of generating instructions for motion to be passed on to subservient movers (*animae*), on the basis of whatever information was physically available to the mind. This function he was coming to see as unnecessary. If the pattern of information suggested the motion in direct enough analogy for a primitive mind to process it, physical mechanisms alone sufficed just as well. The change in Kepler's attitude was a gradual one, occurring over a period of years and never really completed. Celestial minds remained a part of Kepler's universe; only from his astronomy can one see them discreetly withdrawing.

The axis of latitude was not quite in the same alignment over the ages; for the nodes of the planets' latitudinal motion were not fixed, but made their way, with an exceedingly slow motion, around the ecliptic. The direction of the inclined axis varied, then. This fact alone argued that something beyond blind natural forces governed the deviations of the planets north and south of the solar equator. Rather than transfer all the responsibility to a super-corporeal, or mental, influence, Kepler judged it better to retain a quasi-magnetic force, with possible mental supervision, as he had in Chapter 57. To recall his reasoning there (p. 120), the supposition of a mind did not solve the physical problem, but only avoided it. The comprehension which came with a mathematically precise physics was usually imperfect; but it should not on that account be discarded for the more inclusive, but less constraining (and hence less useful), option of a theory of heavenly minds.

A more interesting question arose from the comparison of the axis of latitudinal motion with the axis of libratory force. If each axis, as seemed likely, was a manifestation of some extended structure within the planetary body, then it certainly was desirable that the structures should be the same. The axis of libratory force, as we have seen, was directed perpendicular to the apsidal line, toward the zodiacal locations in which the planet was at mean distance from the sun. The proposed axis of latitude pointed toward the northern and southern limits. Even a cursory examination of the elements of planetary orbits revealed that the limits did not coincide with the mean distances, or indeed bear any obvious relation to them at all. It seemed that the theories of libration and of latitude could not be unified. Two separate axes were required. These were unpleasant straits, for as Kepler had earlier remarked, the analogy of the earth with its axis of rotation made it difficult to see how *any* extended structure inside a planet could maintain a constant orientation, other than that of the rotational axis. Any other axis would trace out a cone around the axis of rotation, and would in the long run act as if it were parallel to that axis. Perhaps, Kepler speculated, the outer shell of a planetary globe was able to rotate freely, leaving the interior undisturbed in its alignment.[144]

Problematic though it was, this example of the earth may have been what set Kepler on the path to the remarkable speculations of Chapter 68, which united the physical theories of latitude and libration after all. The axis of the earth's libratory force pointed at the mean distances, in Aries and Libra. But what of the axis of latitude? The earth's limits of latitude were unknown.

This was a new problem, of course, one which would have made no sense prior to Kepler's physical interpretation of planetary motion. The earth had no motion in latitude, for the earth's orbit defined the ecliptic, the plane of zero latitude. That the choice of this plane was perfectly arbitrary, physically, must have been obvious to Kepler, the first radical Copernican. His physics of motion in latitude made no reference to the plane of the earth's orbit. The latitudinal digressions of a planet carried it away from the plane of motion of the sun's image: the solar equatorial plane. This was the plane in which all the planets would have moved, had they contained no "virtuous diameters" directing their motion north and south. More concretely, this was the plane that was important in the physics. The limits of a planet's motion marked its greatest separations from the plane in which the solar virtue acted.

However nice physically, Kepler's redefinition of latitude had one obvious liability. The solar equatorial plane was unmarked, and observationally invisible. Indeed, the fact of solar rotation was for Kepler only a physical deduction from the circulation of the planets, with no direct observational support. If his physics of latitude was right, then an axis in the body of each planet was aimed at certain places in the heavens. These were the points in the orbit farthest removed from the invisible path of the sun's image. Kepler called this path the kingly circle, *circulus regius*, on account of its importance in his physics. In the absence of evidence regarding solar rotation (sunspots were as yet unknown) or the planetary axes themselves, this conclusion might well have remained sterile. Kepler's irrepressible talent for synthesis combined with a bit of good fortune to develop it into a delightful theory, as misguided as any of his physics yet possessed of genuine scientific elegance.

Tycho Brahe, as it happened, had noticed that the latitudes of some of the "fixed" stars had changed, minutely, from the values catalogued by Ptolemy. Stars near the summer solstice had moved slightly to the north: that is, the latitudes of northern stars had increased slightly, and the latitudes of southern stars had diminished. Near the winter solstice he had observed the opposite,

---

[144] *G. W.*, 3: 392–393.

a small motion to the south. The changes had been tiny, a few minutes of arc, and had decreased to zero around the equinoxes. Kepler had been aware of this phenomenon as early as February of 1599, and had interpreted it helio-centrically by April of that year.[145] In light of the later development of his own physics, Kepler's heliocentric interpretation of Tycho's changing stellar latitudes suggested a boldly speculative unification of his physical theories.

The heliocentric interpretation was simple. The stars could not have moved; Copernican cosmology supposed them to be immovable, at an im-mense distance above the planets. The ecliptic, however, was merely the orbital plane of one of the planets, the earth. If the distance of stars from the ecliptic varied, it was far more reasonable to suppose that the ecliptic moved than the stars. Analogously, the nodes of Mars—the intersections of its plane with that of the earth—had not remained fixed; and on a smaller scale, the lunar nodes moved completely around the zodiac in only nineteen years. In view of these facts, a motion of the earth's orbital plane, the ecliptic, would hardly be surprising, and would go a long way toward explaining many perplexities, both observational and physical.

Figure 28 shows why a motion of the earth's orbital plane accounted for Tycho's observation of changing stellar latitudes. (Incidentally, this explana-tion was in general outline correct.) Two positions of the ecliptic are shown. This plane is inclined at an angle, greatly exaggerated in the figure, to the solar equatorial plane, the mean ecliptic. The nodes, or intersections of the orbital plane with the mean ecliptic, are marked with their traditional symbols: ☊ for the ascending node, ☋ for the descending node. Over the ages the earth's orbit rocks very slowly around the sun, maintaining a constant inclination, so that the nodal line moves from $☊_1☋_1$ to $☊_2☋_2$. As shown, the latitude of a star A near the nodes changes much faster than the latitude of a star B near the limits.[146] Over a very long period, of course, while the nodal line makes a complete circuit, any star's latitude will vary almost sinusoidally as the ecliptic oscillates around its mean position, which is that of the "mean ecliptic" (hence Kepler's term). In the short term, however, and for motions this slow a couple of millennia is a short term, the latitude of a star changes much faster near the nodes than near the limits.

The implications of all this were clear, and must have been exciting to Kepler. Tycho had observed changing stellar latitudes around the solstices, that is, around Cancer and Capricorn. Those regions must therefore contain the terrestrial nodes, the intersection of the ecliptic with the mean ecliptic. The terrestrial limits, where pointed the earth's axis of motion in latitude, thus lay around the equinoxes. But according to Kepler's physics, the earth's axis of libration also pointed near the equinoxes: not because they were the equinoxes, but because, *coincidentally*, they were the mean distances. The terrestrial apsides were near the beginning of Cancer and Capricorn, so the

---

[145] *G. W.*, 3: 403: 10–17; 13: 292, 307.
[146] *G. W.*, 3: 404: 20–39.

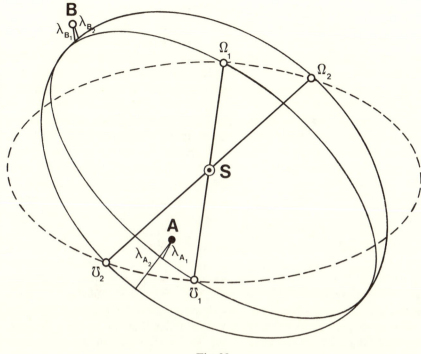

Fig. 28

axis of libration was directed at right angles, near the beginning of Aries and Libra. It could, therefore, be the very same axis as that responsible for the earth's latitudes—and the two physical theories would be united.[147] From Kepler's point of view, the true significance of the changing stellar latitudes which Tycho had observed was not that they were near the solstices, but that they were near the earth's apsides. From such an unexpected source had come evidence uniting two of his theories.

Not one to abandon a trail this promising, Kepler developed further the implications of uniting his theories of libration and of latitude. The chief obstacle was still the invisibility of the mean ecliptic, and hence of his entire latitude theory, and all its parts. Only the orbits were observed; the *circulus regius* or mean ecliptic, to which they inclined, was unmarked. Yet precisely here was where Kepler perceived the practical benefits from his synthesis, tentative though it was. Apsidal lines were not invisible: he could calculate them with some precision. But if the apsidal line of a planet coincided with its nodes, as was suggested by his physics and Tycho's stellar latitudes, then the nodal lines of the planets were known.

All of the nodal lines, of course, were intersections of the various orbits

---

[147] *G. W.*, 3: 404: 40–405: 15.

with the mean ecliptic. All of the nodal lines lay *in* the mean ecliptic. If, Kepler concluded, he could find a plane passing through all of the planetary apsidal lines, that plane would be the hitherto invisible mean ecliptic, marking the equator of the sun's rotation, and the action of its moving image.[148] All of his physics: the rotation of the sun, the axis governing change in distance, and the deviations in latitude controlled by that same axis; all would be spectacularly corroborated.

It is, I hope, superfluous to state again that Kepler's physics was not correct, and that no plane passed through the apsidal lines of the six planets he knew. Evidently he held out some hope that recalculation of the planetary orbits, along the lines of what he had done for Mars and the earth, might yet support the grand synthesis he envisioned. This was not an altogether unreasonable hope; in the *Astronomia nova* Kepler had shifted all the apsidal lines from the mean to the true sun, and had reconstructed latitude theory from top to bottom. The aphelia of the three superior planets were all north of the "true" (terrestrial) ecliptic, and all lay in the semicircle from Capricorn to Cancer. These northern aphelia thus lay between the earth's hypothetical perihelion-node, in Capricorn, and its aphelion-node in Cancer: perhaps, therefore, in a great circle passing through such nodes. This great circle would be the *circulus regius*. If the mean ecliptic lay thus, with the southern limit of the earth's orbit in Libra, then Tycho's changing latitudes implied that the earth's nodes were retrogressing, rather than progressing. Retrograde motion, Kepler pointed out, seemed a more likely direction on the analogy of the lunar nodes, our only clear example, whose motion was retrograde and much faster than that of planetary nodes.[149]

It is worth noticing that Kepler had deduced, from his physical theories, a phenomenon not only important but observable: a plane containing all of the apsidal lines. This plane, had it existed, would have been the solar equator, the fundamental reference plane of physical astronomy. The preeminent position of such a plane was a supposition underlying all of Kepler's physical theories. He did not merely spin out his fancies, however, in the manner of previous theorists of celestial physics, as more-or-less plausible but quite undecidable hypotheses. His own physical theories, although obviously nourished by a fertile imagination, were grounded as firmly as possible in mathematical astronomy. Kepler's tentative identification of the location of the solar equatorial plane with a plane containing the apsidal lines closed the circle, repaying the debt his physics owed to astronomy by deducing the whereabouts of something invisible to that ancient science. The interplay in Chapter 68 among diverse lines of theoretical development, and between all of these and observation, was quite characteristic of Kepler's work.

In this instance, of course, both of his physical theories were wrong, and their briefly glimpsed unity was an illusion. Still, from his reflection on these

---

[148] *G. W.*, 3: 405: 16–22.
[149] *G. W.*, 3: 405: 23–27, 4–6.

matters Kepler gained insight into his (and our) position on a Copernican earth. In the remainder of Chapter 68 he was able to show how the combined effects of the motions of the terrestrial and planetary orbits would produce complex long-term inequalities in the inclination of the Martian orbit to the ecliptic, in the rate of precession, and in the obliquity of the ecliptic. He realized, of course, that his observational data were far too scant for him to quantify this kind of insight. Kepler closed the chapter, one of the most original and beautiful in the entire book, with the hope that it would please God to allow the human race sufficient time on this world to resolve these matters.

Chapter 4

Epitome of Copernican Astronomy

Kepler published the most systematic of his astronomical works, the *Epitome Astronomiae Copernicanae*, in three sections between 1618 and 1621. The *Epitome* was a detailed, if idiosyncratic, textbook of heliocentric astronomy, covering everything from elementary spherical astronomy to Kepler's own discoveries in planetary and lunar theory. It was intended for the student, and written in the form of questions and answers. This book was virtually free of the observational reports and parameter derivations which characterized advanced treatises such as the *Almagest* and *De revolutionibus*. Later, in the mid-seventeenth century, the *Epitome* acquired considerable popularity as a textbook, benefiting from a gradual recognition of Kepler's *Rudolphine Tables* (1627) as the best available.[1]

On the first page of the work Kepler asserted that astronomy is a part of physics. Throughout he maintained a comprehensive viewpoint from which his subject included not only the mathematical description of appearances in the heavens, but also investigations into the archetypal reasons and physical causes of these phenomena. He enumerated five parts of astronomy: observations, geometrical hypotheses, the physical causes underlying the hypotheses, calculations from them, and the instruments used in observing and reproducing the motions. While acknowledging that the study of physical causes was not generally thought necessary to an astronomer, he insisted that astronomers should derive their principles from "a higher science, physics or metaphysics," in order to attain a truer picture of the world.[2]

When he published Book Four in 1620 he felt obliged once again to defend

---

[1] J. L. Russell, "Kepler's Laws of Planetary Motion, 1609–1666," *British Journal for the History of Science* 2 (1964): 1–24.

[2] *G. W.*, 7: 23–25.

his subject matter. The first three books, published together in 1618, had dealt with spherical astronomy, a solid, familiar subject. This solitary fourth book was devoted almost entirely to physics. In the preface Kepler confessed that his book proposed "so many new and unheard-of things concerning all the nature of the heavens, that you could doubt whether you were studying a part of astronomy, or rather of physics; unless you knew that speculative astronomy itself is all a part of physics." [3] This novel claim permeated the *Epitome* from beginning to end: astronomy was physics, and astronomical phenomena were best understood through mathematical study of their physical causes.

The *Epitome* was in seven books, whose material Kepler divided into three sections, not entirely congruent with those in which he published it. The first three books covered spherical astronomy; the fourth through sixth planetary and lunar theory, or *doctrina theorica*; and the seventh precession and related material, traditionally ascribed to the "motion of the eighth sphere," but which pertained in Kepler's view both to spherical and planetary astronomy. The spherical astronomy of the early books was unconventional chiefly in its heliocentric, or Copernican, interpretation of the diurnal rotation of the heavens, and in its account of the likely physical causes of this motion. The later books, however, described Kepler's own theories: elliptical orbits, the area law, orbital planes passing through the center of the sun, and the various archetypal relations and physical forces underlying the structure and dynamics of the universe. In its details, then, the book was no epitome of Copernican astronomy, but rather of Kepler's own. At the beginning of Book Five, wherein it would become evident that Copernicus's specific models were not merely being supplemented, but replaced altogether, Kepler explained why the new hypotheses should still be called Copernican. The elliptical orbits and the rest, he claimed, "arise by necessary physical arguments out of the immobility of the sun and the motion of the earth, Copernican doctrines; and hence they also can rightly be referred to Copernicus." [4] This passage first stated what is probably the best explanation why we think it reasonable, when speaking in general terms, to consider Copernicus's astronomy modern. The heliocentric transformation opened the way for a dynamic physical astronomy centered on the sun. From the dynamics of the solar system modern astronomy has arisen "by necessary physical arguments," even though these arguments themselves have eventually needed to be replaced.

We will not follow the order of Kepler's exposition in discussing the *Epitome*, as we did for the *Astronomia nova*. Instead we shall begin with the planetary theory, drawing material from Books Four, Five, and Six according to our needs. Kepler's mature planetary theory was essentially that developed in the *Astronomia nova*, but here his presentation was more systematic, and his mathematical control of the theory appreciably greater. The improvement will be particularly noticeable at two points. First, Kepler's accumulation

---

[3] *G. W.*, 7: 251: 15–18.
[4] *G. W.*, 7: 365: 1–3.

(in effect, his integration) of the force of libration was done accurately and precisely in the *Epitome*, although the physical theory, harshly clarified by the mathematics, revealed its shortcomings with greater distinctness than before. Second, Kepler was able to clear up completely the relation between the "distance law" and the "area law." That is, he was able at last to justify his use of what we now know as his "second law," on the basis of a physical attenuation of the solar force which he believed to carry planets around the sun.

Also worthy of attention is Kepler's account of his "third law" relating distances of the planets from the sun to their periodic times. This relation was Kepler's principal discovery since his work on Mars. For us his brief discussion of it is both interesting and instructive, because it highlights some of the differences between Keplerian and classical physics, and the traps one can fall into if one is unaware of these differences.

After all of this we shall move on to lunar theory, a subject which had occupied Kepler since the 1590s (as we can see from his correspondence), but about which he had published little prior to the *Epitome*. The theory of the moon was a difficult one, because of the need to explain three additional inequalities implicated in its motion: the "evection," known to Ptolemy; the "variation" discovered recently by Tycho; and the "annual equation." This last, which seems to have been noticed independently by Tycho and Kepler, was not for Kepler an inequality of lunar motion at all, but rather of the earth's diurnal rotation. The study of these inequalities elicited from Kepler a rather elegant elaboration of his basic planetary theory. Our analysis of this theory will take us deeper than we have yet ventured into the archetypal relations which had guided the creation of Kepler's universe. All of this detailed lunar theory remains essentially unknown; it has nowhere been discussed in any depth.

The theory of the moon is easily the most original part of the *Epitome*. It is technically complicated. We shall therefore pause, from time to time, in our discussion of the simpler theory of the major planets, to formulate some of the basic equations of "Kepler motion"—that is, of motion on an ellipse according to the area law. We do this not because such equations are them- selves of historical interest, but because they will furnish a point of reference for our later analysis of lunar theory. The distinctive characteristics of the theory of the moon stand out more clearly in comparison to an explicit Keplerian theory of simple "Kepler motion."

At the beginning of Book Five Kepler summarized his planetary theory, describing the three components of the motion of each primary planet.[5] We are familiar with these components from the *Astronomia nova*. The first was simply revolution about the sun, in a circular motion. If there were no "magnetic fibers" in a planetary body, then "the sun by a rotation of its body on its own axis, carrying an immaterial image of its body through the whole extent of the universe, would carry around the planet in its grasp." Kepler

---

[5] *G. W.*, 7: 363.

carefully specified the regularity of motion under these hypothetical circumstances. If the planet were initially in the ecliptic—by which be meant,
obviously, the mean ecliptic or solar equatorial plane—it would remain there
always. It would return to the same points on every circuit. The orbital center
would be identical with the center of the sun; the orbit would be a perfect
circle; and the planet would be carried with absolutely constant velocity in all
portions of the circle.

Such was not the case; for the planetary body contained two kinds of fibers
(*fibrae*), which deflected it from this simplest of orbits. One kind, remaining
almost parallel throughout the orbit, steered its motion out of the mean
ecliptic into an inclined plane. A slight deflection of these fibers caused the
nodes of this inclined orbit to move slowly in a retrograde direction. The other
kind of fibers, of which one end was attracted by the sun and the other end
repelled by it, made the orbit eccentric to the sun and constricted it slightly
into an ellipse. The planet was thus closer to the sun at some times than at
others. Since the immaterial image of the solar body, which carried the planet
around the sun, was thinner and weaker at greater distances, the planet moved
faster when closer to the sun, and slower when more distant. These fibers also
were deflected slightly from their parallel position, so that the points of
greatest and least distance moved slowly, this time in the direct sense.[6]

Consider first the revolution of the planets around the sun. In Chapter 34
of the *Astronomia nova* (p. 70, above) Kepler had supposed that the solar body
must rotate, and that an immaterial image (*species immateriata*) of the rotating
sun carried the planets around it. In the intervening years the observation of
sunspots moving across the face of the sun had confirmed this supposition.
Kepler believed that a motive soul was necessary to account for the persistence
of the solar rotation. The speed of the rotation was determined by the constant
strength of this motive soul, as compared to the resistance (*contumacia*) of the
matter in the sun's body. He offered several reasons why it should be thought
plausible that there was a soul in the body of the sun, reasons stemming from
his cosmology, whose analysis would lead us far astray. Generally he thought
that a soul was necessary to sustain the rotation, but that no mind or intelligence was needed, because the motion was constant in direction and speed.[7]

The rotating sun carried around with it an image of its own body, which
rotated at the same rate as the body itself. This image was able to grasp the
planets, overcome their natural resistance to motion, and carry them around.
Kepler discussed the magnet as an analogy showing the possibility of such
behavior. Magnets acted only upon bodies similar to themselves, and under
certain circumstances could either attract or repel according to their configuration. By this analogy one pole of the solar "magnet" had to correspond
to the entire surface of the sun, and the other evidently to the center.[8] The
planets could then be ordinary spherical magnets, like Gilbert's *terrelae*; the

---

[6] *G. W.*, 7: 363: 3–37.
[7] *G. W.*, 7: 298–299.
[8] It is customary, but quite senseless, to remark at this point that no such magnet can exist.

pole corresponding to the center of the sun would thus be attracted by the sun's surface, and the opposite pole repelled by it.[9]

This account was essentially the same as that in the *Astronomia nova*, differing only in a more careful attribution of the active role to the sun and the passive to the planet. In this part of the theory, concerning the circumsolar revolution, the difference was primarily one of terminology: the *Epitome* specified more precisely the relation of the action to the acting and acted-upon bodies. I suspect that the motivation for the clarification of roles came from Kepler's libration theory, and particularly from certain changes in it which he had made since the *Astronomia nova*. In the account of circumsolar revolution the active/passive distinction was mentioned only in passing; and indeed it was extraneous to the magnetic analogy, where both the planets and the sun equally must be magnets.

Since the motion of the planets depended upon the sun's image (*species*) being carried around with the rotation of the solar body, Kepler had to explain why the planets did not all have the same period of revolution, namely that in which the sun turned on its axis. There were two reasons. First, the planets, like all matter, exhibited a certain resistance to being moved, which Kepler called among other things *inertia*. They were "inclined, because of matter, to remain in their place."[10] Their velocity at any time was determined by the strength of the force acting upon the planet, relative to this resistance. The conceptual framework here was Aristotelian, although the forces driving Kepler's system were unlike anything in Aristotle. The matter in the planets resisted motion, and since the various planets contained different amounts of matter, the more massive of them moved more slowly, *caeteribus paribus*.

A second reason for differences in speed was the attenuation of the sun's motive force with distance. As we have seen in the *Astronomia nova*, Kepler believed this force to vary inversely with distance from the sun. He carefully justified this relation in the *Epitome*, arguing that it was in fact reasonable for the motive force to follow a different law than light. He based his argument upon a distinction between "the immaterial image of the solar body, flowing out to the planets and beyond," and "its force or energy, which close at hand grasps and moves the planet."[11] The image, although immaterial, was the subject for the force. As it spread through all the universe, the image suffered attenuation just as did light, in the squared proportion of distance from the source. In contrast, the force was not a substance (not even an immaterial substance), but merely an attribute, a virtue or ability, which the solar image possessed when in contact with the body of a planet. In empty space, where the image was spreading out and thinning as it expanded, there was no motive virtue, and hence no question of its weakening. The virtue came into play only

---

[9] *G. W.*, 7: 299: 42 – 300: 40.
[10] *G. W.*, 7: 301: 24.
[11] *G. W.*, 7: 302: 43 – 303: 2.

where the sun's image moved through a planetary body. It acted in one direction only, the direction of the sun's rotation around the zodiac, and it was weakened only insofar as its subject, the moving solar image, was thinned out in this one direction. The motive effect of the sun's image, therefore, decreased in simple inverse proportion to distance from the sun.

This weakening could be observed both in individual planets, which moved more slowly at aphelion than at perihelion, and among the planets, for the more distant ones were slower in their orbits. The effect upon an individual planet was a central pillar of Kepler's physics; we have seen how he developed this notion into the area law. The comparison between planets had become much more interesting shortly before Kepler published the *Epitome*, with his discovery in 1618 of the beautiful relation we now call his third law.

The square of a planet's periodic time was proportional to the cube of its mean distance from the sun. This discovery had solved a question that puzzled Kepler for more than two decades, and had further vindicated his belief in a mathematical order underlying creation. In the *Epitome* he gave a brief account of the physical causes of the planetary orbits. Four distinct factors were involved.[12] First, the circumference of a more distant orbit was longer, in proportion to its radius. Secondly, the sun's power to move a planet weakened following the inverse proportion to distance. For these reasons alone, the periodic times of different planets would have varied as the squares of the orbital radii—if the planets had been identical. Evidently the planets were not identical. As we have said, a planet resisted motion because of the *inertia* of its matter, so that the resistance would be greater, and the period longer, in proportion to the quantity of matter (*copia materiae*), what we would call the mass of the planet. Moreover, a planet that was physically larger experienced the effect of the solar virtue through its whole volume (*moles*). The sun's image acquired its virtue of carrying a planet by contact, as it permeated and moved through the body of the planet; and it acquired more virtue, therefore, in proportion to the volume of the planet. The effect, then, of these two scarcely observable attributes of a planet was to increase the period proportionally to mass, and to decrease it proportionally to volume; and thus to increase the period proportionally to density.

Since the general factors, length of path and strength of force, would together increase the period as the square of distance from the sun, while the actual periods only grew as the 3/2 power of distance, it was clear that the planetary densities must decrease as the square root of distance, to explain the observed relation. Kepler speculated about the component relations connecting distance to mass and volume; in a created universe there need be no stinginess with mathematical harmonies. He had originally believed that, for archetypal reasons, the planets had been formed with surface areas proportional to distance from the sun. However, observations that had been made with the "Belgian telescopic apparatus" had convinced him that Mars could

---

[12] *G. W.*, 7: 306: 37–307: 24.

be but very little larger than the earth.[13] He therefore adopted the less drastic hypothesis that the volumes, rather than the surface areas, were proportional to distance. This forced him to conclude, in a frankly *ad hoc* manner, that the masses of the planets grew as the square root of their distances from the sun.

Obviously, this account of the physics of the third law did not constitute a purely scientific explanation. This alerts us to a distinction which cannot be overemphasized. For Kepler, his "third law" was no law at all, at least not so far as concerned natural science. As Gingerich has observed, it was "not ... a fundamental law in itself, but simply ... a clear and accurate manifestation of the more fundamental principles underlying the cosmos—both physical and archetypal."[14] It was clearly of archetypal importance, and could not have been unintended by the Creator; only in this (important) sense was it a law of nature. In physics it was not a theoretical result (such as it became with the fifteenth proposition of Newton's *Principia*); it was an empirical fact, and nothing more.

The habit of regarding Kepler's period-distance relation as a "law" has hindered efforts to understand its true role in his physics. The elegance of the relation is due particularly to the fact that, as stated, it does not refer to any particular characteristics of the planets involved. This is possible in Newtonian physics because the relation holds for any planet. A planet's inertia is proportional to its mass, as is the gravitational force which overcomes that inertia. The orbit and period are therefore independent of mass, and so it is possible for the third law to avoid reference to attributes of the actual planets.[15] In Kepler's physics there could be no question of this happening. A planet's *inertia* increased with mass, but there was no corresponding increase in the force exerted upon it by the sun. The period-distance relation held for the existing planets only because of the fact (which was contingent physically, though not archetypally) that their densities were such as to produce that relation.

Failure to appreciate the contingency of the period-distance relation has led more than one commentator badly astray. Aiton, for instance, has asserted that the distance law Kepler developed for Mars "could not be extended from one orbit to another, for the periodic times of the planets would then be

---

[13] *G. W.*, 7: 282: 9–24.

[14] O. Gingerich, "The Origins of Kepler's Third Law," in A. and P. Beer, ed., *Kepler: Four Hundred Years*, in *Vistas in Astronomy* (New York: Pergamon, 1975): 18: 600. In a study of the "third law" in the *Harmonice Mundi*, V. Bialas concluded that the relation did indeed have the character of a law of nature for Kepler, but for archetypal (not physical) reasons. V. Bialas, "Die Bedeutung des dritten Planetengesetzes für das Werk von Johannes Kepler," *Philosophia Naturalis*, 13 (1971): 42–55.

[15] We are oversimplifying here, but the above corresponds to the simple idealization where Newtonian dynamics yields Kepler's third law exactly. That is, we neglect the planet's reciprocal effect on the sun, as well as all perturbations from third bodies or lack of spherical symmetry in either the sun or the planet. This is the situation in the *Principia*, I, proposition 15.

proportional to the squares of the distances, a relationship that was false."[16] This would indeed be a problem for identical or arbitrary planets; but Kepler never claimed that the relation would hold for arbitrary planets, only that it did in fact hold for the six real planets. Aiton also misinterpreted Kepler's deductions about the masses and volumes of the planets, thinking that he made "the action of the 'species' ... independent of distance ...," in order to explain the third law. What Kepler said was that the volumes of the planets increased in the same proportion that the motive force weakened; and that *for this reason* all six planets experienced the same force.[17] Their periods therefore varied only because of the lengths of the orbits (in simple proportion to distance) and the resistance of their different masses (in proportion to the square root of distance). That the planets experienced the same force was thus no retraction of the distance law, but a consequence of the circumstances within which the law acted. All of this confusion arose from the question of how Kepler's physics could explain his third law: a moot question, for it was not a law to be explained physically, but rather a simple fact, from which Kepler tentatively deduced some probable relations about the masses and volumes of the planets.

Still more instructive is the argument of U. Hoyer,[18] who recognized all four factors in Kepler's account, discussed how the mass and volume relations were deduced from the period-distance rule, and still was unable to free himself from the preconception that the latter relation was a law of nature, "das dritten Keplersche Gesetz." By assuming that it must hold universally in Kepler's physics, he was of course able to conjure up "inconsistencies." These are of no relevance to anything Kepler wrote, but they do illustrate the perils of bringing modern preceptions to the study of older and fundamentally different sciences.

Finally and most remarkably, H.-J. Treder has inferred (evidently from Kepler's remarks on the coherence of the earth in the preface to the *Astronomia nova*) that the solar forces driving the planets are supposed to be strengthened in proportion to the planetary mass, no doubt by some presentiment of a *Gravitationsgesetz*, although Kepler never said so and although his discussion of the planets cannot be reconciled with this view. Since this alleged theory is incompatible with the rest of Kepler's physics, Treder concludes that with the discovery of the "third law" Kepler's original interpretation of the "second law" was rendered untenable.[19] Treder's tangle of anachronisms, ponderously

---

[16] E. J. Aiton, *The Vortex Theory of Planetary Motion* (New York: American Elsevier, 1972), pp. 18–19.
[17] *G. W.*, 7: 307: 28–42. Kepler explained this to Philipp Müller in letter #938, *G. W.*, 18: 105.
[18] U. Hoyer, "Über die Unvereinbarkeit der drei Keplerschen Gesetze mit der Aristotelischen Mechanik," *Centaurus* 20 (1976): 196–209; "Kepler's Celestial Mechanics," *Vistas in Astronomy* 23 (1979): 69–74.
[19] H.-J. Treder, "Kepler und die Begründung der Dynamik," *Die Sterne* 49 (1973): 44–48; "Die Dynamik der Kreisbewegungen der Himmelskörper und des freien Falls bei Aristoteles, Copernicus, Kepler, und Descartes," *Colloquia Copernicana IV*, vol. 14 of *Studia Copernicana* (Warsaw: Académie Polonaise des Sciences, 1975), pp. 105–150.

encoded in vector equations, conveys an unwitting warning against the un-
critical use of modern concepts in the history of science.

Libration Theory

As we have remarked, the attenuation of the sun's motive virtue between orbits
was far less central to Kepler's astronomical and physical investigations than
its attenuation within a single orbit, between perihelion and aphelion. For a
single planet the physics was much easier to control, the attenuation much
easier to isolate. Yet the weakening of the solar force could influence a planet's
motion only if there was some independent reason why the planet was sometimes closer and sometimes farther from the sun. To account for the change
in distance Kepler retained in the *Epitome* a modified version of his earlier
theory of "libration," by which he meant the planet's alternating approach to
and withdrawal from the sun. Figure 29 shows the theory in outline. The fibers
(of which we show only one, for each of eight positions of the planetary body)
remain pointing in about the same direction, perpendicular to the apsidal line
AE, as the planet moves around the sun. One end of the fibers, here indicated

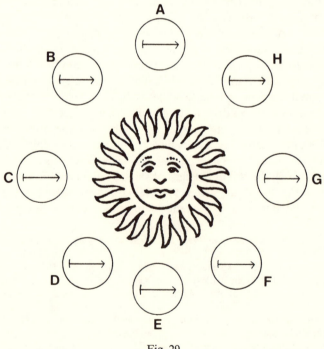

Fig. 29

by an arrowhead, is attracted by the sun; the other end, rounded in the figure, is repelled by it.

At aphelion A each end is exposed equally to the sun; so the sun neither attracts nor repels the planet. However, the sun's grasping force (*vis prensatrix, vis prensandi*) carries the planet around the sun until soon, as at B, the fibers are in a position where they are attracted by the sun more than repelled. The planet therefore is drawn toward the sun. This attraction increases until, somewhere around quadrature C, it reaches its maximum when the fibers are pointed directly at the sun. Thereafter the attraction weakens, although the planet continues to be drawn toward the sun, until at E the attracted end of the fibers is no longer closer to the sun than the repelled end. The planet is swept onward still, its repelled hemisphere now turned to the sun, and hence it is pushed out until it attains its original distance at A. The symmetry of the physics in the descending and ascending halves ensures that the planet must be near A when it is restored to its original distance.[20]

The above account, from Book Four of the *Epitome*, differs from the qualitative libration theory of the *Astronomia nova* in two respects: it is in the passive voice, and it is vague on certain details. First, the passivity: in the *Astronomia nova*, the planet moves toward the sun; in the *Epitome* it is pulled toward the sun. The change, which had already begun in Chapter 57 of the earlier work, does not cause any problems: Kepler had originally argued for an active role for the planets on account of their different eccentricities and apsidal lines, but he could, and did, attribute different strengths and directions to the fibers while changing their role to a passive one.[21] He himself explained this change in his theory with four reasons, fairly plausible reasons; we shall later suggest a fifth one. The first three reasons introduce nothing new: they were either the fruit of further consideration of the *Astronomia nova* theory, or (I think more likely) *post hoc* justifications for a change of mind provoked by the fourth and possibly fifth reasons. These three reasons were that the approach and withdrawal happen radially, directly toward or from the sun, and hence in a direction which was usually not that of the fibers; that the fibers had been required to perform the apparently contradictory tasks of moving the planet and retaining their own direction; and that the enormous sun seemed better capable of extending its reach to the planets, than they of reaching it.[22]

The fourth, and I believe the most important, of Kepler's reasons was also the source of the vagueness in the general account of libration cited above from Book Four. The fibers, he had decided, were deflected slightly from their original direction in the course of the orbit. It was much easier to explain why they should be deflected by the sun than to argue that they rotated slightly of

---

[20] *G. W.*, 7: 337: 8–338: 8.
[21] *G. W.*, 3: 254–256, 349; 7: 338: 40.
[22] *G. W.*, 7: 336: 10–23.

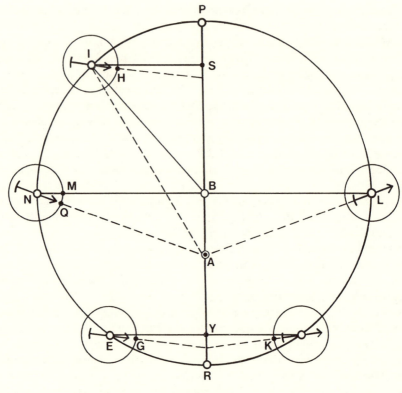

Fig. 30

their own accord, pointing all the while in nearly the same direction and carrying the planet with them in an entirely different direction.

There can be no doubt of Kepler's reason for introducing this (rather messy) deflection[23] into his libration theory: it solved a most vexing mathematical problem. Meanwhile, his physical explanation for the deflection was simple.[24] The sun attracted one end of the fibers and repelled the other; one could well suppose that it was able to twist them slightly away from their proper direction. In Figure 30, the fibers at I have begun to deflect from the original direction IS to the direction IH, a little more toward the sun. If one supposes that they are deflected far enough to point directly at the sun at N, when the planet has completed one-quarter of its orbit, then it is clear that the fibers will thereafter be pulled upward, toward their original direction. Thus with the planet at E, the end G of the fiber is attracted by the sun toward its original alignment along line EY.

---

[23] Kepler's word was *inclinatio*; I have used "deflection" to avoid confusion with latitude theory, and because that word seems best to describe what was happening. *G. W.*, 7: 336: 24–27.
[24] *G. W.*, 7: 339: 8–23.

Since the inclination increased through the first half PN of the descent, it is reasonable that the fiber will exactly recover its original direction in the second half NR. The sun acts at a shorter distance in the lower quadrant; but it has less time to act, since the planet is being carried around faster, and that in the same proportion as the distance. Since the delay is proportional to the distance, and hence to the attenuation of the rotational solar force, this claim is certainly plausible. At R, then, the fiber would again be perpendicular to the apsidal line PR. In the ascending half of the orbit, the repelled end K of the fiber is pushed down, deflecting the fiber until that point L in the orbit when it points directly away from the sun; thence the repelled end is pushed up until aphelion, where the fiber would again be perpendicular to the apsidal line.

This in essence is the physical theory of libration. We have been following Kepler's account in the fourth book, which emphasized the physical causes. In the fifth book Kepler gave mathematical form to his physics. His treatment was carefully thought out and overall quite impressive, for he was able to deduce the elliptical orbit and the area law from the physics sketched in Book Four. Remember that nothing in the *Epitome* is taken from observations; all of the planetary theory formally derives from physics. Yet we shall see that the derivation did not quite succeed. In order to specify a mathematical physics that would move the planets in ellipses, Kepler had to undermine (tacitly) the physical plausibility of the theory. We shall approach this material by first following Kepler's exposition, and only then commenting on the theory as a whole.

Kepler had already specified the circumsolar motion mathematically: its speed varied inversely with distance from the sun. The radial libration was not so simple, for the amount of libration in a small arc of the orbit depended upon the length of the arc, its distance from the sun, the amount of time the planet was in it, and finally the disposition of the magnetic fibers in the planet toward the sun.[25] He dismissed the first factor: more of the forces would be expended in a longer arc than in a shorter, but by considering equal arcs he could neglect this complication. (That Kepler mentioned this factor at all indicates his anxiety about solving what was essentially a problem in integration, using *ad hoc* geometrical arguments based on very small divisions of the orbit.) The next two factors, the time or "delay" in the arc and the planet's distance from the sun, had real effects but canceled one another. The solar force, attractive or repulsive, varied inversely with distance, since in the *Epitome* it is one and the same grasping force as that which carried the planets around the sun. However, the time spent in an arc varied directly with the distance of the arc from the sun, because of the attenuation of circumsolar motion with distance. The weakened libratory force at a greater distance was thus exactly compensated by the extended time in which this force acted.[26]

---

[25] *G. W.*, 7: 365: 26–366: 23.
[26] *G. W.*, 7: 366: 7–13.

In claiming that this compensation held over all the small arcs into which he divided the orbit, Kepler had to look at these arcs from the point of view of the sun, for it was distance from the sun which affected both the strength of the force and the delay of the planet in any given arc. However, his conclusion—that in libration theory the strength of the force and the length of the delay precisely compensated—effectively freed his calculations from the solar point of view. The amount of libration in any arc—and therefore the shape of the orbit—no longer depended on distance from the sun. Of course, the physical equation, the actual acceleration of the planet, remained dependent on this distance: but since it affected libratory motion exactly as circumsolar motion, Kepler could safely neglect it when considering the shape of the orbit, or the optical equation.

This left only the last of the four sources of variation, the disposition of the fibers toward the sun, to analyze mathematically. Analysis of this variation posed difficult problems for Kepler, essentially involving a trigonometric integration, but as we shall see he solved these problems. The relation he postulated between the disposition of the fibers and the momentary strength of the force "admitted" or "received" in that disposition was simple: the force was proportional to the cosine of the angle between the fibers and the direction from the planet to the sun. This was a reasonable hypothesis, already developed in Chapter 57 of the *Astronomia nova* (p. 113, above). Kepler justified it here with various arguments, one of which (Figure 31) is interesting as a connection with his lunar theory. The sun illuminates half of a planet, leaving the other half in darkness. The two hemispheres are delimited by what Kepler

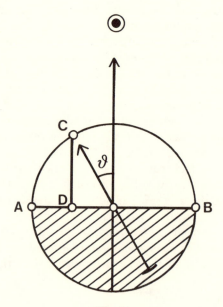

Fig. 31

called the "circle of illumination" (seen edge-on as diameter AB in the figure). We shall later find the circle of illumination playing a prominent role in the theory of the lunar inequalities. At this point[27] Kepler observed that a perpendicular from the plane of this circle to the end of one of the fibers of libration followed the same proportion as the solar force "admitted" in this configuration. Clearly CD in the figure varies as cos Θ, where Θ is the angle between the sun and the fiber. It seemed fitting that the fibers should experience the sun's force according to their distance from the dark hemisphere.

However plausible this cosine is as a measure of the effective libratory force, it is not obvious that one can conveniently measure it. In the course of the orbit, the fibers are deflected from their original direction perpendicular to the apsidal line. To allow for this deflection in his calculations, Kepler had to specify precisely how big it was. His treatment of this question was divided between Books Four and Five,[28] and the specification shifts twice in the course of discussion, supposedly for the sake of clarity. Careful analysis reveals, however, that the theory as a whole is inconsistent. We shall demonstrate this with a rather careful exposition of Kepler's own analysis, for the failure of this component of his libration theory is perhaps the most serious technical shortcoming in Kepler's physical astronomy.[29]

His initial explanation, in Book Four, implied without actually stating that the sine of the deflection was proportional to the sine of the eccentric anomaly. This made it easy to see how, in Figure 30, the deflection was greatest precisely at the quadrant, where the eccentric anomaly was 90°. This was an important consideration because the deflection, having gradually increased over the first quarter of the orbit, had available another quarter, exactly, to return to zero. (The deflection had to be zero after half the orbit. The apse, the place where the solar distance was at a momentary extremum, by definition occurred where the fibers were perpendicular to the solar radius. If the restitution of the fibers took noticeably more or less than half the orbit, the apsidal line would shift correspondingly.) In Book Five he amended the proportionality, "compelled by physical speculation" (*vi speculationis physicae*), saying now that the sine of the deflection was proportional to the sine of the *coequated* anomaly.

It does not take long to work through Kepler's demonstration. In Figure 30,

$$\frac{\sin HIS}{\sin QNB} = \frac{\sin IAP}{\sin NAP} \tag{26}$$

Moreover, by the law of sines,

---

[27] *G. W.*, 7: 367: 19–27.

[28] *G. W.*, 7: 339–341, 369–370.

[29] In what follows, it is helpful to notice that the figure printed on p. 369 of *G. W.*, vol. 7, is not the one cited in the accompanying text. That figure repeats the figure on p. 366; but the text demands the one from p. 339.

$$\frac{BI}{BA} = \frac{\sin BAI}{\sin BIA}$$

$$\frac{BI}{BA} = \frac{\sin IAP}{\sin BIA} \tag{27}$$

Assuming, for the time being, that the orbit is circular, we also have from the law of sines:

$$\frac{BI}{BA} = \frac{BN}{BA} = \frac{\sin BAN}{\sin BNA}$$

$$\frac{BI}{BA} = \frac{\sin NAP}{\sin BNA} \tag{28}$$

Combining now (27) and (28),

$$\frac{\sin IAP}{\sin BIA} = \frac{BI}{BA} = \frac{\sin NAP}{\sin BNA}$$

Finally, we transpose the proportion:

$$\frac{\sin IAP}{\sin NAP} = \frac{\sin BIA}{\sin BNA} \tag{29}$$

In words: the sine of the optical equation is proportional to the sine of the true or coequated anomaly, on a circular orbit. This is really nothing but a simple consequence of the law of sines. Since Kepler claims that the sine of the deflection is proportional to the sine of the true anomaly, equation (26), then also

$$\frac{\sin HIS}{\sin QNB} = \frac{\sin BIA}{\sin BNA} \tag{30}$$

The sine of the deflection is proportional to the sine of the optical equation. It has already been stated that the fibers point at the sun at N (still in Figure 30), so that there the deflection is the optical equation, and hence the proportion is one of equality. That is, angles QNB and BNA are equal. Therefore also

$$HIS = BIA \tag{31}$$

the deflection equals the optical equation. Subtracting these equal angles from angle AIS,

$$AIH = BIS$$

The angle between the solar radius and the magnetic fibers equals the complement of the eccentric anomaly. Taking cosines,

$$\cos AIH = \cos BIS = \sin SBI \tag{32}$$

Since the strength of the solar attraction or repulsion is proportional to the cosine of angle AIH, the angle between the fibers and the sun, the strength is also proportional to sin SBI, the sine of the eccentric anomaly.

This is too good to be true, or at least too good to be accidental. In the *Astronomia nova* (see p. 115) Kepler had struggled with the question why the planet's velocity of approach to the sun appeared to vary as the sine of the eccentric anomaly, although physically it seemed that it should vary as the sine of the true anomaly (since the latter sine equaled the cosine of the angle between the fibers and the sun). The deflection had resolved this question in the most direct way possible. Kepler had simply allowed his fibers to deflect, by the precise amount of the difference between the eccentric and true anomalies: the optical equation of center.

The business about the sine of the deflection being proportional to the sine of the true anomaly was a red herring. In a circular orbit the sine of the optical equation, and hence of Kepler's intended deflection, is indeed proportional to the sine of the true anomaly: we have seen the demonstration. When he moved to an elliptical orbit, Kepler retained, without comment, the proportionality of the attractive force to the sine of the eccentric anomaly, as expressed by (32). This meant that he was using (31) as the measure of the fibers' deflection; and on an elliptical orbit (31) is not equivalent to his original statement (26), that the sine of the deflection is proportional to the sine of the true anomaly. Once again—and tacitly this time—he had changed the rule governing the fiber deflection.

The deflection law given by (31) was thus an arbitrary specification introduced to make the theory work. As we shall see, it made the theory work. However, it was no longer plausible physically. The physical theory of deflection supposed that after the planet passed point N in Figure 30, the solar pull on the magnetic fibers would begin to restore them to their original position, perpendicular to the apsidal line. The deflection therefore reaches its maximum at point N, and decreases thereafter. But the optical equation, which must be the measure of the deflection if the rest of the theory is to work (that is, if the orbit is to be an ellipse), continues to increase after point N. The fibers, to be sure, do point at the sun precisely at the quadrant of the orbit. Thereafter, until the true anomaly reaches 90°, the physical theory has the sun pulling them back toward right angles with the apsidal line, while the necessary mathematical relation (31) requires that they continue to deflect, inexplicably turning against the pull of the sun. This is the case both on a circular orbit, such as Kepler used in his specific discussion of the fiber deflection, and on the elliptical orbit that he introduced after further physical and geometrical discussion.

Kepler's theory of the deflection of the magnetic fibers was thus burdened with a serious contradiction. The cause he postulated for the deflection was the tug of the sun on the attracted end of the magnetic fibers, and its push at the repelled end. The tugging and pushing should have twisted the fibers in a clockwise direction in Figure 30, until the motion of the planet around the sun carried the fibers to a position where they pointed "below" the sun, that is toward perihelion; thereafter the fibers should have been twisted back counterclockwise. The mathematical requirement that the *effect* of the deflection should be a force proportional to the sine of the eccentric anomaly was incompatible with this supposed physical cause; for this mathematical condition implied that the deflection equaled the optical equation, so that the fibers pointed directly at the sun when the eccentric anomaly was 90°. After that point, they should physically have been twisted back counterclockwise, but the mathematical condition of equation (31) increased their deflection still further, against the pull of the sun.

Kepler cannot have been unaware of this problem. He knew very well, and had known when writing Chapter 57 of the *Astronomia nova*, that he needed a libratory force proportional to the sine of the eccentric anomaly, although the central role of the sun in his physics seemed to require that he use the true anomaly instead. Once he had hit upon the idea of letting the fibers be deflected, the necessary amount of deflection would have been obvious: the difference between the eccentric and true anomalies. His use, in Part V of the *Epitome*, of a deflection proportional to the sine of the true anomaly was equivalent to the 'correct' deflection law, since he was still discussing circular orbits. (Indeed, a deflection whose sine is proportional to the sine of the true anomaly shares the physical absurdity of the 'correct,' optical-equation deflection, since it obviously continues to increase until the true anomaly reaches 90°.) This alternate formulation was a bit disingenuous, since it capitalized on Kepler's frequent (and frequently plausible) remarks concerning the role of sines in measuring the force of natural phenomena, as exemplified by the scales and the lever. The deflection law in its proper form, equal to the optical equation, would hardly have been so believable to his readers.

After introducing the ellipse, of course, Kepler could only have discussed the deflection in its optical-equation form; instead, he made no mention of the deflection as such, but simply used the libration law as established from (32): the force of libration, at any point in the orbit, was proportional to the sine of the eccentric anomaly. This formulation ignored the inverse dependence on distance from the sun; but Kepler could do this because, as he correctly put it, the weakening of the solar force with distance was exactly compensated by the greater delay of the planet at longer distances.

The unstated[30] exact form of the deflection theory—that the deflection equaled the optical equation—has evidently not been noticed in the literature on Kepler. The whole physical theory of libration, presented didactically in various simplified forms throughout the *Epitome*, seems paradoxically to have eluded the attention of historians. We may note at this point that when Kepler had been discussing the physical basis of the deflection, in Book Four, he had implied that it was proportional to the sine of the eccentric anomaly. This was

---

[30] In an entirely different context, Kepler did mention that the deflection of the libratory fibers equals the optical equation: *G. W.*, 7: 348: 6–7.

credible physically, but would not have produced the correct libration. I think we must recognize that Kepler was concealing a defect in his theory. In a book intended to teach astronomy, and containing many things that were new and surprising, but true, such equivocation may perhaps be justified. We can only wish that Kepler had written a definitive treatise laying out his theories, problems and all, with both the candor of the *Astronomia nova* and the systematic organization of the *Epitome*.

As noted above, Kepler respecified the amount of fiber deflection for the second time, implicitly, when he went on to show that his libration theory yielded not a circular, but an elliptical orbit. It is not surprising that he failed to remark the change. On an ellipse the necessary deflection, governed by the optical equation, was even less plausible than it had been on a circle. The elliptical eccentric anomaly is a peculiar angle indeed, not involving the planet directly at all. If ever Kepler—reverting to his earlier and most abstract mode of physical analysis—tried to understand how a planetary mind could "see" what angle to deflect the fibers, he must have been greatly perplexed. In fact there is no speculation in the *Epitome* about planetary minds, or even souls, in connection with libration theory. With the development of such an elaborate (and generally successful) theory, Kepler perhaps no longer felt the need for the abstract physical critique he had employed in the absence of positive theory. At any rate, it is a fact that the more sophisticated of his physical theories were by this time independent of mental or "animal" assistance. Only the primordial rotations of the sun and the planets were attributed to the operation of souls.

Kepler had shown, by this point, how a seemingly plausible mechanism, the deflection of the planetary fibers under the pull of the solar force, could result in their "admitting" a force proportional to the sine of the eccentric anomaly. His account of how to sum the increments of libration caused by this force was much improved over that in the *Astronomia nova* (compare pp. 114 ff., above). The new account was based on a theorem from Pappus (*Mathematical Collections*, Book Five, Prop. 36): if a sphere is cut by any number of planes perpendicular to a given diameter, that diameter is divided in the same proportions as the surface of the sphere. We may easily see that this is true (Figure 32). Let B be the center of a unit sphere. Cut the sphere by two planes CC′ and DD′, intersecting BP perpendicularly at X_1 and X_2. The area of a slice CDD′C′ of the spherical surface is then given by the integral

$$CDD'C' = \int_{\beta_1}^{\beta_2} (2\pi \sin \beta)\, d\beta$$

$$= -2\pi \cos \beta \Big|_{\beta=\beta_1}^{\beta_2} = 2\pi(\cos \beta 1 - \cos \beta 2)$$

$$= 2\pi(X_1 - X_2)$$

which is obviously proportional to $(X_1 - X_2)$, the length of that section of the diameter cut off by the two planes. Taking this result from Pappus, Kepler

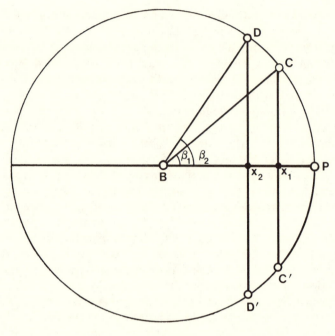

Fig. 32

reasoned as follows. Each increment of libration was proportional to sin β, and the sines were distances such as CX_1 or DX_2 on Figure 32. These distances are the radii of circles on the surface of the sphere, circles centered on the diameter BP; they are therefore proportional to the circumferences of such circles. Now if the surface of the sphere is divided into an infinite number of parallel zones, each zone, without any finite thickness, will simply be a circle. As circles, their proportions to one another will be as their radii CX_1, DX_2, etc. Therefore the sum of any number of such zone-circles varies as the sum of their respective radii, and conversely.

Consider now the physical problem. The sun attracts or repels the planet with a force which varies as the sine of the eccentric anomaly: that is, as the radius of a circle (or infinitely thin zone) such as we considered in the previous paragraph. An accumulation of these increments of force thus varies as the sum of the radii, and hence as the sum of the circumferences, of the zones. As the planet is swept around from C to D, in Figure 32, the infinitely thin zones accumulate to make up the finite zone CDD'C'. By Pappus's theorem the area of this zone is proportional to the segment X_1X_2, a segment which is simply the difference between the cosines of the eccentric anomalies at C and D. If we start at aphelion P, we find that the increments of libration sum to an amount proportional to the spherical surface PCC' when the planet has reached C. Again, by the theorem of Pappus, this is proportional to PX_1, the versed sine of the eccentric anomaly. (By definition vers β = 1 − cos β.)

The cumulative libration is proportional to the versed sine of the eccentric anomaly.

Since the greatest amount of libration has been accomplished at perihelion, where the planet has approached the sun by twice the eccentricity, and since vers $180° = 2$, the constant of proportionality is simply the eccentricity, and we have

$$\text{Libration} = e \text{ vers } \beta \tag{33}$$

and

$$R = \text{aphelial distance} - \text{libration}$$

$$R = (1 + e) - (e \text{ vers } \beta)$$

$$R = 1 + e \cos \beta \tag{34}$$

Kepler could now construct the orbit. Drawing an auxiliary circle PDR (Figure 33), he divided the descending semicircle into equal parts, say at K, G, D, N, S, and dropped the perpendiculars KX, GF, DB, NA, SY to the apsidal line PR. Point A is the sun.[31] Next, he constructed the amounts of libration completed at the given eccentric anomalies PK, PG, etc., in accordance with (33). Specifically, he constructed points M, I, etc., such that

$$\frac{PM}{2 \cdot AB} = \frac{PX}{PR}$$

$$\frac{PI}{2 \cdot AB} = \frac{PF}{PR}, \quad \text{and so on.}$$

Here $2 \cdot AB$ is the maximum libration, and PR is the maximum versed sine to which it corresponds. He had now the amounts of libration PM, PI, PF, PQ, PV, PB, corresponding to the eccentric anomalies measured by arcs PK, PG, PD, PN, PS, PR. That is, he had the amounts by which the planet had approached the sun, from its initial distance PA, at those eccentric anomalies. The resulting distances from the sun were, respectively, AM, AI, AF, AQ, AV, AB. Kepler transferred these distances out to the perpendiculars, to points L, H, E, O, T, R, and concluded that these were the positions of the planet at the stated eccentric anomalies. We shall later take up the question of why the distances were transferred to the perpendiculars KX, GF, etc., instead of, say, to radii BK, BG, etc. (a familiar question from the *Astronomia nova*, as on pp. 124 ff., above).

The auxiliary circle centered on B was thus used only for defining the eccentric anomaly and computing the amounts of libration. It was not needed to show what was happening physically. The planet at aphelion (Figure 34)

---

[31] In Kepler's figure he took, for the sake of convenience, an eccentricity $AB = \frac{1}{2}$ and a division into 30° arcs, so that the perpendicular from N falls on A, and $AB = AR$. Neither of these facts enters into the discussion, except to simplify the diagram and confuse the reader.

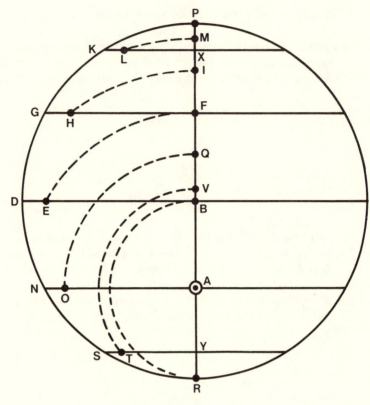

Fig. 33

was distant from the sun by AP. If there had been no magnetic fibers, no libration, the planet would simply have been swept around in the circle PR′ centered on the sun at A. Since the planet did contain fibers it was attracted by the sun, so that by the time it had been moved around through arc PC′ it was closer to the sun by a distance C′C, and was therefore found at C. After half a revolution the planet had been pulled all the way in from R′, where it would have been, to R. The sun then began to push the planet away. After it had been swept through arc RF′, it had been pushed out by F′F, and so was at F. After being swept around full circle, the planet had been pushed out by the same distance P′P = R′R it was earlier drawn in; and so was found where it started, at P.

Ignoring, as we have been, the inverse-distance dependence of the solar force, this was Kepler's physics of planetary motion in longitude. The auxiliary circle (dotted in Figure 34) was formally nothing more than a computing device to take account of the deflection of the magnetic fibers. (Historically, of course, it was the circular orbit Kepler had employed before determining that an oval was needed, and it remained computationally indispensable; but

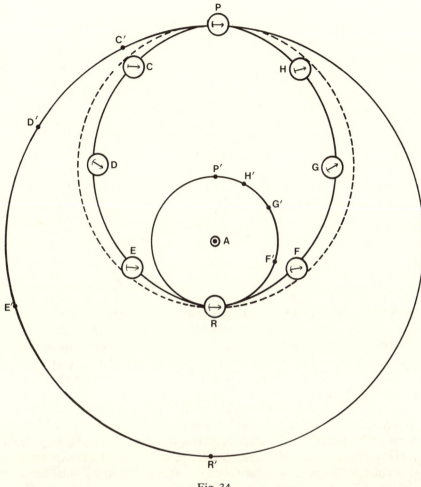

Fig. 34

it had nothing to do with the physics of the orbit.) If there had been no deflection the force of libration would have been measured by the sine of the true anomaly (called it γ); the total libration at any true anomaly γ would have been e vers γ; the distance from the sun would have been 1 + e cos γ; and the orbit would have been, well, different. The function of the auxiliary circle in computing the libration could then have been served by any circle centered on the sun, such as PR'.

Kepler had next to show that the orbit he had constructed, more-or-less from physical theory, was in fact an ellipse. This is a side issue for our present concerns, but we will summarize the argument for the sake of completeness. An ellipse and its foci may be identified by the following two properties. First, if perpendiculars are drawn to the major axis from the circumscribing circle

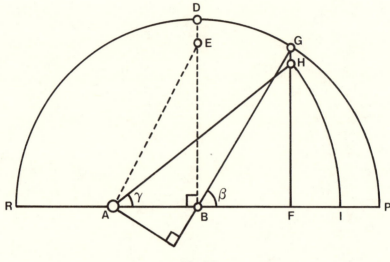

Fig. 35

(as KX, GF, etc., in Figure 33), the ellipse divides them all in the same proportion. Second, two points exist on the major axis which are distant from the ends of the minor axis by the semimajor axis. These points are called the foci; their combined distances from any point on the ellipse equals the major axis. To show that Kepler's construction has these properties we shall use the simplified Figure 35.

Let PGR be half a unit circle with center B, and let A be the sun, and AB = e the eccentricity. For any eccentric anomaly β = PBG, construct I on the diameter PR such that PI = e vers β, or, equivalently, AI = 1 + e cos β. Draw arc IH centered on A and intersecting GF at H. Kepler's theory yields point H, a point on the perpendicular GF to the apsidal line whose solar distance is given by (34). We must show that H lies on an ellipse inscribed in the circle, with one focus at A. To do this, we prove that the ratio HF/GF is constant for any eccentric anomaly β. Indeed,

$$HF^2 = AH^2 - AF^2$$

$$= (1 + e \cos \beta)^2 - (e + \cos \beta)^2$$

After some simple manipulations,

$$HF^2 = (1 - e^2) \sin^2 \beta = (1 - e^2)GF^2$$

$$\frac{HF}{GF} = \sqrt{(1 - e^2)}$$

Since $\sqrt{(1 - e^2)}$ is independent of β, we conclude that the set of all points constructed as H forms an ellipse, with PR the major axis. The minor axis

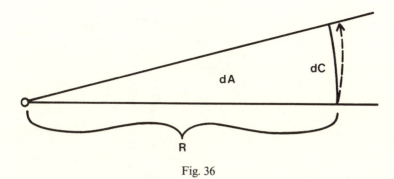

Fig. 36

intercepts the ellipse at a point E such that PBE = β = 90°; but then AE = 1 + cos 90° = 1. Therefore the sun A is a focus of the ellipse.

There remained the area law to derive. Formally, Kepler proved in the *Epitome* that the area law followed from a revised statement of his distance law; the latter was taken as a physical postulate. The distance law as used here stated[32] that the component of the planet's velocity which was perpendicular to the solar radius, that is, the component which actually carried the planet around the sun (rather than toward or away from it), was inversely proportional to distance from the sun. This formulation is clearly equivalent to the area law: if we define C as the component of motion perpendicular to the radius R from the sun (Figure 36), then we have

$$dA = \frac{R \cdot dC}{2}$$

$$\frac{dA}{dt} = \frac{1}{2} \cdot R \cdot \frac{dC}{dt}$$

If the left side of the equation, the rate at which area is swept out by the radius vector, is to be constant, then clearly dC/dt must be inversely proportional to R.

Kepler's demonstration was necessarily less concise. Because of the importance of the result in his physics, and because it is often not realized that he restated the distance law 'correctly' in the *Epitome*, we shall work through it. It is clever in places, but as he presented it a few critical steps were missing. For the sake of clarity we shall diverge somewhat from the details of his argument to fill in these steps, retaining, however, his overall strategy.

Let PORS (Figure 37) be the auxiliary circle, circumscribing the elliptical orbit PCRG, with PR the apsidal line or major axis, B the center, A the sun. Divide the circle into equal arcs, say PO, OQ, etc., at aphelion, and RS, ST, etc., at perihelion, and suppose these arcs to be extremely small. Dropping

---

[32] *G. W.*, 7: 377: 27–43.

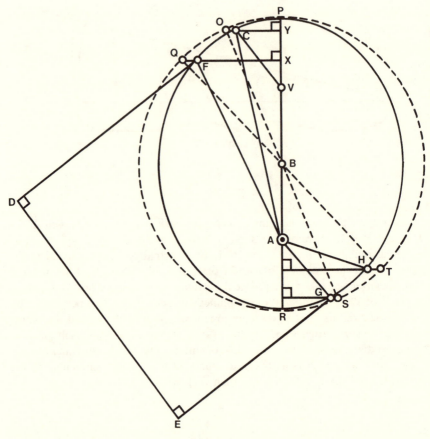

Fig. 37

perpendiculars to the apsidal line from the points of division, we obtain the corresponding points C, F, ..., G, H, ..., on the ellipse. We wish to show that the time required to traverse the elliptical arcs PC, CF, ..., RG, GH, ..., is proportional to the areas of the elliptical sectors from the sun PCA, CFA, ..., RGA, GHA, ..., on the basis of the restated distance law.

We first prove that all of the elliptical sectors from the center B have the same area. The circular sectors POB, OQB, ..., obviously have the same area, since they were constructed to equal arcs. But since the ellipse divides the perpendiculars OY, QX, ... in a constant ratio $YC/YO = XF/XQ = \sqrt{(1 - e^2)}$, the area of an elliptical sector is equal to $\sqrt{(1 - e^2)}$ times the area of the corresponding circular sector. Hence the elliptical sectors PCB, CFB, ..., RGB, GHB, ..., are all equal.

Next, consider the elliptical arcs in pairs that are opposite one another (with respect to the center B): thus PC and RG, CF and GH. The sum of areas of the two sectors from the center to any such pair equals the sum of the sectors from the sun to that pair. This is obvious only for the sectors on the apsidal line, PCB and RGB. This pair of sectors may be considered as right triangles; and their bases PC and RG are equal by symmetry. So

$$PCB + RGB = \tfrac{1}{2} \cdot PC \cdot PB + \tfrac{1}{2} \cdot RG \cdot RB$$

$$= \tfrac{1}{2} \cdot PC \cdot (PB + RB) = \tfrac{1}{2} \cdot PC \cdot (PA + RA)$$

$$= \tfrac{1}{2} \cdot PC \cdot PA + \tfrac{1}{2} \cdot RG \cdot RA$$

$$= PCA + RGA$$

This result—that the sum of the two sectors from the center to a pair of opposite elliptical arcs equals the sum of the sectors from the focus to these arcs—is also valid away from the apsidal line, although this becomes clear only after a bit of discussion. Again we consider the opposing tiny arcs CF, GH, as line segments.[33] By the symmetry of their construction they are equal and parallel. As before, the sum of the sectors, or triangles, to these two arcs is simply half the product of their length with the perpendicular distance between them. In the figure we have extended the arcs (taken as line segments) to show clearly the perpendicular distance DE between them. Regardless of whether A or B is taken as the vertex, the area of a pair of triangles based on the given arcs is simply half of DE times the arc length. Since we have already established that all of the sectors from B are equal, we conclude that all of the sums of opposite sectors from A are equal.

Now that we have established some of the characteristics of these elliptical sectors which are determined by perpendiculars dropped to the apsidal line from equal divisions of the auxiliary circle, let us consider the delays of the planet in traversing the tiny sections of the orbit corresponding to these sectors. According to the restated distance law, the times or delays will be proportional to distance from the sun, for arcs whose components perpendicular to the radius from the sun are the same. The question immediately arises whether this is the case for the arcs we have been considering. To show that it is indeed the case—that if these elliptical arcs are decomposed into components directed *at* the sun and components directed *around* it, the components around the sun will all be equal—we shall use the simplified Figure 38.

The divisions of the ellipse in this figure are as before. What we have done is to decompose the planet's motion from C to F into a circumsolar component CM due to the sun's rotating image or *species*, and a radial component MF which is the libration; and likewise for the opposite arc, GH. Kepler performed just such a decomposition.[34] We shall show that the circumsolar components are equal: CM = GN.

Draw CV, connecting C to the empty forcus of the ellipse. A well-known

---

[33] Kepler called these sectors "triangula ... (rectilinea vel quasi)," *G. W.*, 7: 377: 8–9. The argument which follows, generalizing the equality of paired central sectors to paired focal sectors, is not made explicit in the *Epitome*. It is relatively simple, however, except in the awkwardness of the figure.

[34] *G. W.*, 7: 377: 36–41.

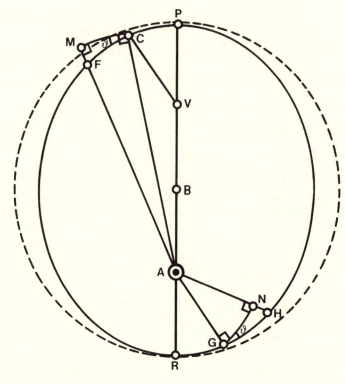

Fig. 38

property of an ellipse is that lines from the two foci to any point on the ellipse make equal angles with the tangent to the ellipse at that point. Thus AC and VC make equal angles with the tangent to the ellipse at C. But (by symmetry) the angle between VC and the tangent at C is the same as the angle between AG and the tangent at G. Substituting, the angle between AC and the tangent at C equals the angle between AG and the tangent at G. The complements of these angles—FCM and HGN—are therefore equal: say FCM = HGN = Θ. Since the elliptical arcs CF and GH are equal, we have immediately:

$$CM = CF \cos \Theta = GH \cos \Theta = GN$$

Stripped of its libration, the planet moves the same distance around the sun in the tiny arc CF near aphelion as in the opposite arc GH near perihelion.

In fact, the amount of circumsolar motion is the same in all of the elliptical arcs thus constructed; for

$$\text{area CFA} + \text{area GHA} = \tfrac{1}{2} \cdot CM \cdot (AC + AG)$$
$$= \tfrac{1}{2} \cdot CM \cdot (AC + VC), \quad \text{by symmetry}$$
$$= \tfrac{1}{2} \cdot CM \cdot PR,$$

because the sum of distances from the foci to any point on an ellipse equals the major axis. Thus

$$CM = \frac{2 \cdot (\text{area CFA} + \text{area GHA})}{PR}$$

But the sum of the areas is a constant for all pairs of opposite arcs. Thus the amount of circumsolar motion is the same for each and every arc.[35]

The distance law therefore applies directly to the arcs which were constructed by dropping perpendiculars to the apsidal line from the equally-divided auxiliary circle. The delay in each arc is proportional to its distance from the sun. But the same proportion holds among the areas of the sectors, for their bases perpendicular to the solar radii (namely the arcs PC, CM, ..., RG, GN, etc.) are all equal. Therefore the areas of the sectors are proportional to the delays in the arcs. By restating his distance law, Kepler had finally justified his use of the area law as a substitute for it.[36]

Kepler continued his exposition, as he had in the *Astronomia nova* (p. 130), with the practical observation that one could compute areas on the circle much more easily, and that these were entirely equivalent to the elliptical areas. In Figure 39, the time elapsed to eccentric anomaly β was properly measured by the elliptical sector PHA. Because of the property of ellipses used earlier—that an ellipse cuts all the perpendiculars GF, DB, etc., in the ratio $\sqrt{(1 - e^2)}$ to 1—we have

$$\text{area PHA} = \text{area PHF} + \text{area HFA}$$

$$= \text{area PGF}\sqrt{(1 - e^2)} + \text{area GFA}\sqrt{(1 - e^2)}$$

$$= \text{area PGA}\sqrt{(1 - e^2)}$$

The elliptical sector is always proportional to the area projected out onto the auxiliary circle from the apsidal line. The latter area is exceedingly easy to compute, for it is the sum of PGB and GBA. PGB is directly proportional to β, and

$$\text{area GBA} = \tfrac{1}{2} \cdot \text{BA} \cdot \text{GF}$$

$$= \tfrac{1}{2} \cdot \text{BA} \cdot (\text{DB} \sin \beta)$$

$$= \text{area DBA} \cdot \sin \beta$$

---

[35] We should note that, although the empty focus V does appear on Kepler's diagram, he did not present the above argument in general, but considered only the arcs at the apsides and mean distances.

[36] Koyré, in his thoughtful but often confused book *The Astronomical Revolution*, pp. 320–322, objected to Caspar's alternate demonstration of this result, on the grounds that the action of the circumsolar force was essentially curvilinear, and that therefore Caspar should not have approximated it with straight line segments. Kepler, however, was perfectly capable himself of using line segments to approximate circular arcs—as in this very passage (*G. W.*, 7: 377: 7–20). Caspar's demonstration is not mistaken or unfaithful; only superfluous.

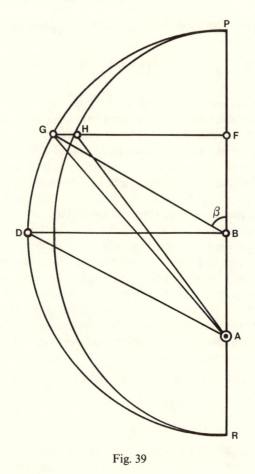

Fig. 39

One can simply calculate the maximum triangle DBA $= \frac{1}{2} \cdot$ e, once and for all, and any of the triangles GBA corresponding to the physical equation is instantly available as DBA $\cdot$ sin β.

Let us compute, for future reference, the mean anomaly on the ellipse according to the area law. Still in Figure 39, the mean anomaly at H is proportional to area PHA, and hence to area PGA. Where the radius PB $= 1$, and AB $=$ e, the area of the circle is π; and

$$PGA = PGB + GBA$$

$$= \frac{\beta}{2\pi}(\pi) + \frac{1}{2} \cdot AB \cdot GF$$

$$= \frac{1}{2} \cdot (\beta + e \sin \beta)$$

In order to convert area PGA to the pseudoangle α of mean anomaly, we must norm it by the condition that the area of the whole circle (π, in area measure) correspond to the circumference of the circle (2π, in angular measure). This norm requires that the angular measure of the mean anomaly α be twice the area measure of sector PGA. Thus

$$\alpha = \beta + e \sin \beta \tag{35}$$

This is, of course, "Kepler's equation," in the form where angles are measured from aphelion.

This completes the derivation of Kepler's physical theory of planetary motion in longitude. Let us pause to consider some of the peculiarities of the theory, characteristics which Kepler naturally did not emphasize in his own account. Most striking is the strange definition of the eccentric anomaly β, an angle which is very convenient mathematically (it played a central role in the later development of celestial mechanics) but which appears to have no physical significance whatsoever. Geometrically, β is defined to be the angle, as seen from the center of the orbit, between aphelion and the projection perpendicular to the apsidal line of the planet's position onto the auxiliary circle. The angle can hardly acquire any physical significance from this defini-tion; as we have already remarked, the auxiliary circle and the center of the orbit are entirely without physical meaning. Yet β is the free variable in the mathematical description of the orbit. Surely there must be some way in which this mysterious angle is determined by Kepler's physics.

The question really has two parts. The role of β in (35), "Kepler's equation," which expresses his distance law, is quite natural. Because of the equivalence between the areas of elliptical sectors and areas of circular sectors, β is convenient for measuring the area of elliptical sectors; and these, in Kepler's physics, are important because they express the cumulative effect of the distance law. But the areas measure the cumulative effect of the distance law only because the orbit is an ellipse,[37] and we have yet to determine what role the peculiar angle β played in the physics leading to the ellipse.

The eccentric anomaly β formally entered our account of this physics in equation (32), which stated that the physically-important cosine of the angle between the inclined magnetic fibers and the radius from the sun equals the sine of β. This equation was derived *before* the shape of the orbit was specified. Three components went into equation (32): the original position of the fibers, a physical postulate; the deflection of the fibers, another physical postulate; and the eccentric anomaly, which at that point appeared to retain its 'natural' or conventional signification, namely the angle at the center of the orbit between the planet and aphelion. This signification turned out to be illusory. After the law (34) governing the distance of the planet from the sun had been derived on the basis of (32), Kepler placed the planet not on a leg of the angle β, but on a perpendicular dropped to the apsidal line from the intersection of

---

[37] Actually because the orbit is a conic section: the area law and the properly-stated distance law are equally compatible with the results of Newtonian dynamics.

that leg with the auxiliary circle. β was revealed as a different angle than one would have thought.

In our account of the *Astronomia nova* we discussed the adventures of the eccentric anomaly at some length: how its meaning became ambiguous when Kepler discarded the circular orbit; how he was nevertheless unable to do without it, and so employed it in one sense for his distance law and in another for his libration theory; and how he finally recognized that a completely new interpretation would reconcile the equations of Mars with the distances, on an elegantly elliptical orbit. We are now going to forget all that for a few pages, to take Kepler's physics seriously and clarify the meaning of the eccentric anomaly in the libration theory leading to the ellipse.

In the physical analysis leading to the fundamental equation (32) governing the libration, the eccentric anomaly is arrived at by deflecting the fibers through some small angle from their initial position. Specifically, if the true anomaly is γ and the deflection of the fibers is δ (measured in the clockwise direction), then

$$\beta = \gamma + \delta \tag{36}$$

Equation (36) is mathematically equivalent to (32), and to our earlier conclusion that the deflection equals the optical equation. Physically, however, the deflection δ is a meaningful angle determined by some kind of interaction between the sun and the planet—and it is Kepler's failure to give any consistent account of this interaction which defeats his physics.

To see how the fiber deflection determines the shape of the orbit, *via* the eccentric anomaly β as defined by (36), consider Figures 40 and 41. In Figure 40 we have three possible positions of the planet, corresponding to three

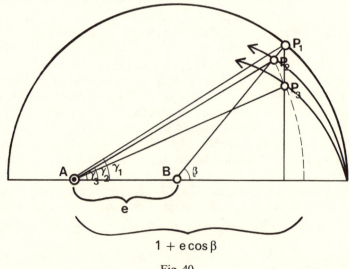

Fig. 40

different orbits, all for a given eccentric anomaly β. The first position P_1 (not consistent with Kepler's libration theory) is that of an eccentric circular orbit, with β the central angle to the planet. The other two positions are both compatible with everything Kepler said explicitly about physics. They lie, therefore, on the dotted arc at radius $R = 1 + e \cos \beta$ from the sun. The path defined by points P_2 is the *via buccosa* of Chapter 58 of the *Astronomia nova*, for which β is still the central angle to the planet. P_3, of course, is on the ellipse, and for it β has its new, surprising definition. Physically, the difference between the shapes of these two orbits must come from different libration rates, and hence from different laws governing the deflection of the magnetic fibers. (The orbits differ also, of course, in the speed at which the planet would move. On the *via buccosa*, the physical distance law cannot be reduced to Kepler's equation.)

The positions in Figure 40 correspond to the same "eccentric anomaly," variously defined, but to different locations of the planet in its course around the sun. We can see more readily the effects of the different laws of fiber deflection by comparing them at the same *true* anomaly, as in Figure 41. Here the planet is located, according to the different theories, for a specific angle γ of true anomaly. As before, P_1 is on the eccentric circle. P_2, the *via buccosa* position, is on the leg of the angle of its eccentric anomaly, β_2; while P_3, the ellipse position, lies on the perpendicular dropped from where its eccentric anomaly β_3 cuts the circle. Evidently $\beta_3 > \beta_2$, else the two positions would not lie on the same line AP_3P_2 of true anomaly. Since the same distance law, $R = 1 + e \cos \beta$, was used in constructing both the *via buccosa* and the ellipse,

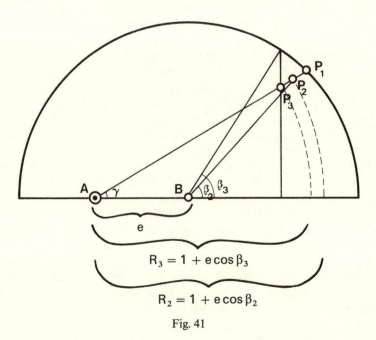

Fig. 41

we know that $R_3 < R_2$, as shown in the figure. The ellipse lies inside the *via buccosa*. Physically, there can be only one reason for this. To produce an elliptical orbit, the fibers in the planetary body must deflect by a greater angle than they would if the *via buccosa* were the orbit. By deflecting farther, they point more nearly at the sun, and admit more of its libratory virtue; and so the planet's librations are greater in the mean distances.

We have arrived at this conclusion by comparing the respective values β_2 and β_3 of eccentric anomaly which the two theories require to yield the same true anomaly. We could do this because we knew in advance the geometrical interpretations of β which go with each of the theories. However, if we restrict ourselves to Kepler's physics, β is defined only as a consequence of the fiber deflection, by equation (36). From this point of view the differing amounts of deflection take causal precedence. Moreover, we cannot explain physically the difference in deflection between the two theories, for Kepler gave us no consistent account. We have seen that the mathematics of libration theory require a definite amount of deflection—equation (31)—in order to predict the correct distance—equation (34)—but when confronted with more than one theory that uses this distance law, we do not have enough physical theory to determine which deflection law is correct.

In these last pages we have been leaning on Kepler's celestial physics to see where it bends. The flexible point in this theory was the deflection of the magnetic fibers of libration. He gave two laws for this deflection and implicitly used a third. The first ($\sin \delta = e \sin \beta$) was physically plausible; the latter two were not. Furthermore, it is clear that once the deflection was separated from any specific causal mechanism, the shape of the orbit was no longer determined physically. Kepler's physics was not adequate to support his "first law."

Quite the opposite may be said of his "second law," which had chronologically been his first. From its initial, flawed formulation Kepler had based his distance law on physical principles; and by all evidence the area law, in the *Epitome* as in the *Astronomia nova*, was never anything more than a computational technique. The earlier form of the distance law, indeed, was 'incorrect,' that is, not equivalent to the area law. Yet I think it is fair to say that the later form, wherein delay is proportional to distance in an arc of given length as measured *perpendicular to the solar radius*, is an altogether more natural conclusion from Kepler's physical principles. He had always treated the planet's orbit, in its plane, as the result of two forces, one of which we have termed circumsolar and the other librational. It was only the former which he believed to vary simply as the inverse of distance. It may be, as Aiton has argued, that Kepler originally thought that the libratory mechanism "simply steered the planet, while the 'species immateriata' alone pushed it along the eccentric circular orbit."[38] At any rate Kepler implied, knowingly or not, that this was the division of labor. Yet as early as Chapter 33 of the *Astronomia nova* Kepler's phrasing had been that "however much farther the planet is

---

[38] Aiton, "Infinitesimals," p. 294.

from that point which is taken for the center of the universe, so much weaker is it impelled around that point (*circa illud punctum*).[39] Obviously this statement does not necessarily imply the final form of the distance law, with the motion decomposed into components around and toward the sun; but just as obviously it is entirely consistent with that final form.

To understand the transition between the two forms of the distance law, I think we must take into account the enormous increase in the sophistication of Kepler's physical thought over the course of his career. When the distance law first appeared in his writings, in Chapter 22 of the *Mysterium* (p. 18 above), his conceptual apparatus was of the simplest sort: the planet was about as much faster than its mean motion, at perihelion, as it was slower at aphelion. In Chapter 32 of the *Astronomia nova* (p. 62), where he demonstrated the equivalence of the equant to his distance law, the physical argument was still at about the same level, although the mathematics had been made explicit. In subsequent chapters of that book Kepler pursued the libratory force which determined the shape of the orbit, first in exceedingly abstract terms and then, increasingly, as a distinct component of his theory, to be isolated and pinned down with numbers. By Chapters 59–60 (p. 129) he was so far from the crude concept of speed being inversely proportional to distance that he emphasized the unequal lengths of the arcs to which the distance law applied.[40] Yet he did not restate the fundamental distance law. In its naive form—speed decreased in proportion to distance from the sun—this relation had been with him so long, and had served him so well, that he did not yet realize how the details of his theory had, subtly, altered the circumstances in which it operated.

And so the distance law had retained its original formulation in the *Astronomia nova*, as Kepler's physical analysis developed from its intuitive beginnings into a fairly sophisticated mathematical treatment of two independent forces, one of which carried the planet around the sun, and the other toward and away from it. In the ensuing decade Kepler finally achieved an orthogonal decomposition of the motion, and realized that to compute the equations he must apply his distance law to the circumsolar motion alone, abstracted from the libration. When we view it in the long course of the development of Kepler's physical analysis, the explicit reformulation of the distance law constitutes more a correction of overly-simple mathematics than a revision of fundamental theory. The fundamental notions had been revised gradually, all along, by the force of the physics Kepler had constructed upon them.

Before leaving Kepler's account of the motion of primary planets in longitude, we should note that the libratory force, like the circumsolar force, was indeed attenuated in proportion to distance from the sun. This fact was implicit in Kepler's treatment of libration, which he considered in relation simply to the planet's progress around the orbit—its eccentric anomaly—and not in relation to time. The dependence of eccentric anomaly upon distance therefore applied also to the libration. As Kepler put it, the planet experienced

---

[39] *G. W.*, 3: 236: 12–13.
[40] As Aiton, indeed, has observed; "Infinitesimals," p. 303.

a weaker libratory force when farther from the sun, but this weakening was compensated by the planet's longer delay in an arc at the greater distance. Although he did not dwell upon the attenuation of the libratory force with distance, there can be no doubt that Kepler was aware of it.[41]

Latitude in the Epitome

Kepler's account of motion in latitude, in contrast to that of the motions in longitude and distance, was clean and simple. It was essentially unchanged from the *Astronomia nova* (pp. 130 ff., above). The body of the planet contained fibers which Kepler described as analogous to an steering oar (*remus, temo, gubernaculum*), although their function was more that of a keel. They deflected the planet in the direction toward which they pointed, slightly north or south of the ecliptic, away from the direct line along which the solar image, like wind on a ship's sails, was pushing it.

Within the context of his physics, this was a simple enough explanation for a simple phenomenon. Since Kepler lacked a concept of inertia as continuing motion, he needed something to move the planets out of the mean ecliptic. Whatever this was, it had to be specific to each planet, since each had its own nodes and limits. Some sort of fibrous structure was the obvious choice to explain motion within a plane. Of course, the planets were beginning to have a rather complex internal structure (the fibers of libration had to progress slowly, those of latitude to retrogress slowly) which one would not have suspected from external appearances, but this could not be helped.

The fiber hypothesis did permit an explanation of the nodes' retrogression, which Kepler put forward with some hesitation, as he was not certain of the phenomenon itself in the primary planets. In Figure 42, the fibers of latitude within the planetary body are shown at the limits A, C, and at the nodes B, D. Suppose that the solar image, moving counterclockwise about the sun, exerts pressure on these fibers, and moreover that the pressure is stronger near the sun. Then at the limits A and C the pressure will be greater on the end of the fibers closer to the sun; the fibers will therefore be twisted very slightly clockwise, that is *in antecedentia*. The nodes and limits will follow the direction of the fibers.[42]

Summary: Planetary Theory in the Epitome

We conclude our account of planetary theory in the *Epitome* by observing that with all its faults, it was probably about as good as Kepler could have done in the absence of a proper law of inertia. His physics was a physics of

---

[41] See, for instance, *G. W.*, 7: 365: 33; also the letter to P. Müller of 1622, No 938 in *G. W.*, 18: 107, lines 243–246.

[42] *G. W.*, 7: 347: 22–39.

Fig. 42

velocity, as Newton's was a physics of acceleration. His two forces, taken together, are at every point of the orbit proportional to the velocity vector of a planet in what we call "Kepler motion": that is, an elliptical orbit traversed in accordance with the area law. One can express his theories in vector notation, differentiate with respect to time, and show that they imply an acceleration directed toward the sun, and inversely proportional to the square of distance from it. Kepler tuned his physics to velocity instead; and a planet's velocity is by no means as simple as its acceleration. In the natural physical coordinates, distance R and true anomaly γ,

$$\text{Libration} \sim \frac{\sin \gamma}{R \cdot (1 - e \cos \gamma)} \tag{37}$$

Had he tried, Kepler would have been hard-pressed to concoct a simple, plausible force that varied as required by (37). As we have seen, he used the much simpler expression of the libration in terms of the eccentric anomaly β,

$$\text{Libration} \sim \frac{\sin \beta}{R} \tag{38}$$

Although β is a convenient parameter for describing the orbit geometrically, Kepler could introduce it into the natural set of physical coordinates only by an arbitrary deflection of the fibers of libration. As we have seen, it was this deflection, induced by the necessity of bringing β into the physics somehow, which defeated his physical theory.

Before turning to the special problems and delights of Kepler's lunar theory, let us glean what we can of the fate of his bold attempt, described in Chapter 68 of the *Astronomia nova*, to identify the fibers causing motion in latitude with those responsible for the libration (p.133 above). He had worked out this theory, speculatively, on the basis of minute changes in the observed latitude of certain fixed stars. At the time, he had probably not been able to test it, because he needed the apsidal lines and nodes of all the planets, calculated precisely and with respect to the true sun. Toward the end of the *Epitome* there is some discussion which bears on this theory, indirectly but unmistakably.

The seventh and final book of the *Epitome*, whose subject matter Kepler described ironically as the "motion of the eighth and ninth spheres," was almost entirely devoid of physics. He discussed therein the precession of the equinoxes, the changes in the direction and magnitude of the earth's eccentricity and in the obliquity of the equator, and the significance of the "mean ecliptic" or solar equatorial plane, which he had introduced in Chapter 68 of the *Astronomia nova*: topics which (excepting the last) had indeed been explained in pre-Copernican astronomy by postulating particular motions of the eighth sphere, that of the fixed stars, and of others exterior to it. Kepler's account, which as always took full advantage of the heliocentric viewpoint, correctly attributed these phenomena to very slow motions of the earth's axis. He repeated his earlier suggestion that deviations from constancy in the earth's rotation implied some sort of mental supervision for the animal faculty that rotated the earth; but did not elaborate.

In the *Astronomia nova* Kepler had indicated how such considerations might simplify his planetary physics. The physical nodes of the planetary orbits were properly their intersections with the solar equator or mean ecliptic, rather than the true ecliptic. If these nodes turned out to coincide with the apsidal lines, a single set of fibers for each planet, perpendicular to the nodal/apsidal line, would suffice for both libration and motion in latitude. By

Table 3. Kepler's apsidal lines

Planet	Longitude of aphelion	Latitude of aphelion
Saturn	Sagittarius 26°	1; 4° N
Jupiter	Libra 7°	1;20° N
Mars	Leo 29°	1;48° N
Venus	Aquarius 2°	2;34° S
Mercury	Sagittarius 15°	3;37° N

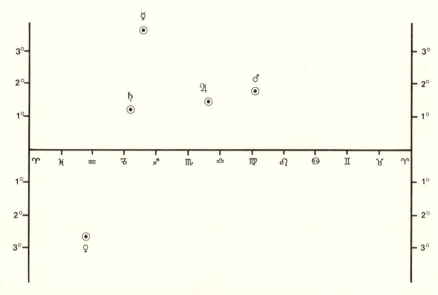

Fig. 43

the time Book Seven of the *Epitome* was published, Kepler had computed the elements of all the orbits and could check his supposition. Like most bold predictions from incorrect theory, it failed.

We can reconstruct the downfall of Kepler's attempt to unite his theories of distance (or "altitude") and of latitude, using the values he gave in Book Six for the apsidal lines, inclinations, and nodal lines of the planetary orbits with respect to the true ecliptic.[43] From these we can compute the heliocentric latitudes of the aphelia (see Table 3). In Figure 43 we have graphed the aphelia onto the rolled-out zodiac. If they all lay in a plane, their graph would appear very nearly as a sine curve. Evidently it does not. In Book Seven Kepler simply remarked that the aphelia of all the planets approximately marked the mean ecliptic, but differed from it somewhat, without commenting explicitly on his earlier theory.[44]

Epitome: Lunar Theory

Let us now look at Kepler's lunar theory. It is evident in his correspondence that from his early years he had devoted a great deal of attention to the special problems of lunar theory. By 1599 he had identified (independently of Tycho,

---

[43] *G. W.,* 7: 416, 427, 433, 438–439.
[44] *G. W.,* 7: 520: 24–26.

it seems) the fourth equation of lunar motion, now known as the "annual equation," although he did not believe that it pertained to the moon itself.[45] In February of 1601 he outlined early theories of the lunar inequalities in a letter to Maestlin.[46] He withheld his theoretical speculations, however, until the publication of the *Epitome*, by which time he had developed the elaborate physical theory needed to account for all of the equations. As we have already remarked, this entire theory is virtually unknown today. We shall consider it in some detail.

The lunar theory is contained mostly within Books Four and Six of the *Epitome*. It differs from Kepler's planetary theory in two ways. First, he had to allow for the apparent fact that not only the sun, but much lesser bodies such as the earth could have all the physical apparatus of the central body in a planetary system: the rotating image and the power of attracting and repelling the appropriate ends of magnetic fibers. Further, he had to explain the second, third, and fourth inequalities observed in lunar motion, none of which had any known analogue in the motion of primary planets.

In the first book of the *Epitome*, Kepler had discussed the earth's rotation at some length, primarily, it seems, because this rotation was an important tenet of Copernican cosmology, while at the same time being inherently implausible. His explanation of the diurnal rotation of the earth was naturally the same as that of the solar rotation. The earth's body contained circular magnetic fibers parallel to its equator. By means of these fibers the earth turned itself in that direction toward which the fibers were directed, from west to east as it happened. Kepler suspected that a soul (*anima*) was the actual cause of the motion, and the circular fibers its instruments. This soul supplied the force by which the earth's material *inertia* was overcome.

Kepler did not think it at all strange to attribute a soul to the earth. As further indications of the earth's quasi-animal nature he remarked on the subterranean heat; the earth's production of minerals, rivers, and vaporous exhalations; the spontaneous generation of small animals; and the geometrical shapes found in crystalline minerals. He was not certain whether the slow precession of the earth's axis implied a mind at work behind the motive soul.[47]

A consequence of the earth's diurnal rotation, in exact analogy to the solar rotation, was that the moon was swept around the earth from west to east. However, the moon's orbit lay always near the ecliptic, not near the equator as it seemed that it should. Kepler attempted to account for this by pointing out that the moon was actually orbiting the sun, albeit erratically, and that it never moved backwards in this solar orbit, not even at new moon when the

---

[45] For an account of Kepler's early work on lunar theory see C. Anschütz, "Über die entdeckung der Variation und der jährlichen Gleichung des Mondes," *Zeitschrift für Mathematik und Physik, Historisch-Literarisch Abtheilung* 31 (1886): 161–171, 201–219; 32 (1887): 1–15.

[46] Letter #183, in *G. W.*, 14: 161–165. In this letter Kepler has the moon on an oval orbit, except when the lunar apogee is at right angles to the sun! This was over a year before he concluded, for entirely different reasons, that the orbit of Mars was an oval.

[47] *G. W.*, 7: 92: 4–42; 91: 39–41.

terrestrial motive image was sweeping it directly backwards. Hence the solar image was clearly dominant, and it was not surprising that the moon's path was approximately in the ecliptic. The attempt here to consider lunar motion with respect not to the earth, as had always been done, but to the solar system itself, was quite proper. It is not clear, however, just why the disturbances caused by the earth in the moon's (solar) orbit would be confined to the neighborhood of the ecliptic. The problem is a difficult one; moons of other planets are indeed more often found near their equatorial planes.

A further problem arose from this: namely that sometimes, near the solstitial colures, the moon moved parallel to the equator and hence parallel to the circular fibers, while at other places, near the equinoxes, the lunar motion was at an appreciable angle to them. Yet the moon moved no faster in Cancer or Capricorn than in Aries or Libra. To this Kepler observed that the moon was further from the equator when moving parallel to it, while its motion was angled to the equator when near it. If the earth's magnetic image was weaker away from the equator, as seemed plausible, then the angle and the intensity would mutually compensate. He was not willing to say that the compensation was perfect, since there were still discrepancies between lunar observations and theory.[48]

Lunar theory was complicated principally by the fact that the sun affected the moon's motion. Both the second and third lunar inequalities (the evection and the variation, as we now know them) patently depended upon the position of the sun. Kepler accounted for this evident influence by supposing that sunlight, as it stimulated the growth of things on earth, also stimulated both the earth's rotation and the movement of the moon around the earth. The consequences of this twofold excitation were not simple, but Kepler had carefully worked them out. He analyzed the lunar motion in longitude into four components: first, the regular motion due to the earth's rotating magnetic image; second, the inequality due to the libratory fibers, as in the primary planets; third, a synodic irregularity in this last inequality; and fourth, a synodic variation in the regular motion itself, called simply the variation. Finally, one had to take into account a small inequality, with a period of an anomalistic year, which Kepler attributed not to the lunar motion but to the terrestrial rotation, and hence to measurements of apparent time. This last is today known as the annual equation of the moon's motion. An additional synodic component also entered into the moon's motion in latitude.

The first inequality in longitude and the first component of motion in latitude were identical to the corresponding phenomena in the motion of primary planets. Kepler called them *inaequalitates solutae*, the independent or separated inequalities, since they could be calculated without regard to the sun. The only problem he had in transferring to the moon all of his libration and latitude theory was the requirement that the magnetic fibers of libration and of latitude remain almost constant in direction through the orbit. This

---

[48] *G. W.*, 7: 320: 37–46.

had been something of a problem for primary planets; if anything it was even more troublesome for the moon, which obviously kept the same face toward the earth.

Kepler had two suggestions for dealing with the problem. It might be that the fibers were not pointed directly toward or away from the earth at mean distance, but were only inclined, by a visually imperceptible amount, from their normal, apsidal position perpendicular to the earth-moon radius. He did not attempt to develop mathematically this greatly modified libration theory, nor have I, but I suspect the attempt would be quite difficult. Nor did he develop the analogous explanation for the fibers of latitude, whose constant direction seems almost requisite if the orbit is to be planar. However, the small angle of inclination, some five degrees, makes it more likely here that the motion of the moon's face in maintaining the direction of the fibers of latitude would go unnoticed.

The other alternative was the rather awkward one to which he had been forced by the contrary directions of motion of the planetary apsides and nodes: that the fibers were imbedded in separate interior shells, which could move independently so as to keep their proper orientations. Neither of these explanations was entirely satisfactory, although Kepler remained confident that some similar explanation would work. "For among these and similar ones," he admitted, "[it is] uncertain what exactly would be the manner of this motion; this alone is certain, that whatever the manner is, it is accomodated to physical and magnetic causes, that is, corporeal, and thus geometrical...."[49]

As mentioned above, Kepler attributed the extra lunar inequalities to the effect of sunlight. He was not very specific about how sunlight was able to stimulate the moon's motion (how could he have been?), but chose sunlight because it came from the right direction, given the observed inequalities, and because it was obviously of some efficacy here on earth. The force he was looking for could not have been the sun's magnetic image, since the lunar motion was always accelerated at syzygy, whether the moon was going in the same or the opposite direction as the solar image. Had he not chosen light as the agent for the synodic lunar inequalities, he would have had to postulate a similar but unobserved agent. The mathematics would have been the same, of course.

How, then, did sunlight affect the motion of the moon? "No part of celestial physics was more difficult than this to explain."[50] In Figure 44, E is the earth and around it are various positions A, B, C, D of the moon. Everything is illuminated by the sun at S. Kepler supposed, for lack of a better idea,[51] that the light of the sun had some faculty of strengthening the motive image of the earth, which swept the moon around in its orbit; and that the efficacy of this faculty depended upon the way in which it was applied (*applicatio*) to the

---

[49] *G. W.*, 7: 350: 29–32.
[50] *G. W.*, 7: 322: 33.
[51] "Cum igitur alia causa non appareat...." *G. W.*, 7: 322: 44.

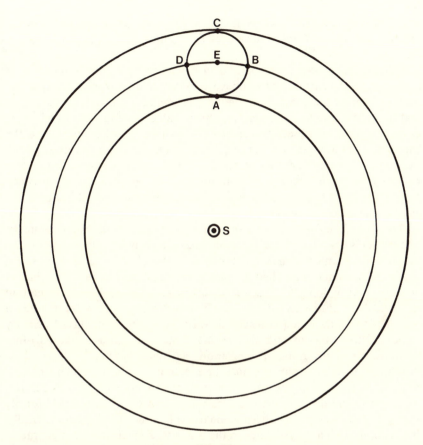

Fig. 44

earth's image. The angle at which a sphere of light, centered on the sun, intersected a sphere of motive virtue, centered on the earth, determined how much the light would intensify the virtue at the point of intersection. Due to the great distance of the sun, a sphere of solar light was scarcely curved at all in regions near the earth. Thus at syzygies, with the moon at A or at C, the two spheres virtually coincided, and the effect was maximal. At quadratures B or D, the spheres were perpendicular and there was no effect at all. Kepler initially gave no physical explanation for variation in intermediate places, where, as it turned out, the effect would vary as the squared cosine of the 'angle of application,' or the squared sine of the angle from quadrature. (The *angle of application* was the angle between the moon's motion and the sphere of sunlight. Since the latter was locally almost a plane, perpendicular to the radius from the sun, the angle of application was very nearly equal to the angle from syzygy to the moon's position in its orbit.) The phenomenon, Kepler explained, was as if the moon were more disposed to move along the surface of a sphere of light than to cross it; or as if the path itself were made easier at the syzygies, harder at the quadratures, like motion with or against the grain on a wooden table.[52]

It was hard to say, from the known inequalities, whether sunlight both strengthened the earth's motive image at syzygy and weakened it at quadrature; or whether the image alone had the weak effect observed at quadrature, and sunlight strengthened it first to its mean value, and finally to its full vigor at syzygy. As we shall see, Kepler eventually chose the latter alternative: the sun could only strengthen, never weaken, the motive image of the earth. Moreover, as the sphere of light was extended in both longitude and latitude, so it hastened the moon's motion in both directions. The close analogy between Kepler's account of the synodic inequalities of longitude and latitude was an elegant feature of his lunar theory.

The moon, then, had two sources of motion, the terrestrial magnetic image and the sun's light. Its motion therefore had twice as many components as the motion of a primary planet, which had only the solar image as a source of motion. A primary planet had its regular or mean motion; the moon had in addition a semimonthly variation in this mean motion. A primary had an inequality due to the eccentricity and elliptical shape of its orbit; the moon had in addition the evection, a synodic variation of this inequality. A primary wandered from the ecliptic; the moon's motion in latitude, like its other motions, varied according to its position relative to the sun.[53]

One of the extra lunar motions, the evection, had been known since the time of Ptolemy, who had explained it geometrically with a (considerable) variation in the distance between the moon and the earth. Since Ptolemy various other attempts had been made to explain the evection geometrically, but without such a large change in the distance. Kepler now claimed, rather proudly, to have accounted for the evection without introducing *any* new mechanism which changed the lunar distance; because his explanation was not geometrical but entirely physical. He needed "no new circles at all, beyond those we are accustomed to use in the demonstration of the separated [inequality], and also in all the primary planets...."[54] He had not arrived at this explanation easily, for his early attempts had involved a lunar eccentricity which varied with the elongation of the moon from the sun. Analysis of eclipses had finally convinced him that the moon's eccentricity was no greater when the apsidal line was at conjunction with the sun than elsewhere, and that, therefore, the synodic inequalities were purely physical, without an optical component.[55] The moon was speeding up and slowing down in the course of its orbit, depending upon how far it was from syzygy. The shape of the orbit was unaffected.

The evection is a periodic change in the size of that equation or irregularity

---

[52] *G. W.*, 7: 323: 7–14.
[53] *G. W.*, 7: 349: 7–46.
[54] *G. W.*, 7: 451: 34–35.
[55] Kepler to Maestlin, letter #884 in *G. W.*, 18: 14.

which was attributed to the lunar epicycle in earlier astronomy, and to the elliptical shape of the orbit by Kepler. Kepler called that other equation, which the moon shared with the primary planets, the separated (*soluta*) inequality. The evection altered its magnitude in the following way. When the lunar apogee pointed directly toward or away from the sun, the moon was always further from its mean position, in the same direction, than it would have been at the same mean anomaly, if apogee had been at quadrature to the sun. That is, the separated inequality was greater when the line of syzygies was near the apsidal line. Kepler's physical explanation for the evection was based on a facile analogy between the earth itself, as the source of motion responsible for the separated inequalities, and the circle of illumination centered on the sun, as the source of motion responsible for the synodic inequalities.

> For while [the moon], by the simple and always uniform law of the eccentricity, goes around the earth its mover, just as each of the primary planets goes around the sun, it happens by accident that it is distant by varying amounts from the other source of its motion, which accelerates it in the syzygies. For if its longer interval from the earth occurs in the syzygies, where the acceleration is maximal, then the image of the earth, being spread out in a more extensive sphere, is weakened in one of the syzygies; and this not only in its native and archetypal vigor, but also in its strengthening acquired from the sun. On the other hand, if the longer interval between the moon and earth coincides with the quadratures, where there is no acceleration, then of the acquired vigor, which is zero, there will also be no loss, and no gain in the short perigeal interval.[56]

Kepler was arguing that the strengthening of the terrestrial image by sunlight varied, like the unstrengthened image, with distance from its source. The obscurity of the argument was due to the separation of this second equation from the intimately related third equation, the variation. Physically, sunlight always accelerated the moon in the syzygies, but by different amounts; for the sake of mathematical convenience Kepler defined the variation as the mean effect of this acceleration. The second inequality measured the irregularity of the acceleration, as explained in the above passage. If the moon was at apogee, farthest from the earth, when it came into line with the sun and the earth, then the extra velocity from sunlight was less than its mean value when in that line. Since the variation would correct for the mean value, a further correction (the evection) was needed to retard the calculated motion. On the other hand, when the moon was at perigee, the variation would understate its actual acceleration, so that the evection would again be needed. Of course, if the syzygies occurred at mean distance, then the variation would be the only correction needed. The evection, then, was a correction to the variation in Kepler's physics. Let us turn to Book Six to see precisely how he developed the mathematics of all this.

---

[56] *G. W.*, 7: 351: 5–14.

In Figure 45, A is the earth and the large eccentric circle is the moon's orbit. (Kepler neglects the elliptical shape of the orbit in computing the evection. The resulting error, as we shall see, is safely negligible.) B is the center of the orbit; hence D is apogee and F perigee. Lines HG and PQ are directed at the sun, and are essentially parallel. The syzygy which is closest to the regular apogee D, in this figure H, is called the monthly apogee; the other syzygy, here G, is called the monthly perigee.[57] HG may thus be called the monthly apsidal line. The 'monthly eccentricity' AC is obtained by projecting the regular eccentricity AB onto the line of syzygies, or the monthly apsidal line, HG.[58] We have labeled the angle DBP = DAH between the regular and monthly apsidal lines (in other words, the elongation of the apsidal line) as κ. The angle κ is slowly changing, of course, because the apparent motion of the sun *in consequentia* is about nine times as fast as the motion of the apsidal line, also *in consequentia*. For convenience, however, Kepler computed the elongation κ of the apsidal line for the moment of time with which he was concerned, and assumed this configuration to be fixed. When he considered the synodic lunar inequalities, Kepler used the term 'technical month' (*mensis technicus*) to express his assumption of a hypothetical month in which the apsidal elongation κ remained constant.[59]

Physically, the line IAK, perpendicular to the line of syzygies, is the circle of illumination on the earth's surface, projected outward onto the lunar orbit. It represents the surface of a sphere of light centered on the sun. Because this small portion of the sphere is essentially without curvature, IAK appears as a line. The monthly eccentricity AC or BZ is the distance of the plane of this circle of illumination, considered as a source of motion, from the center of the orbit; just as the regular eccentricity is the distance of the chief source of motion from the center.[60]

The evection, or as Kepler called it the recurrent monthly inequality (*inaequalitas menstrua temporanea*) was measured by a triangle with the monthly eccentricity as base, just as the *inaequalitas soluta* was measured by a triangle with the regular or *soluta* eccentricity as base. Thus, with the moon at L in Figure 45, triangle ACL measures the evection. Like all areas, ACL is a measure of time or mean anomaly. The sign of the equation ACL is determined analogously to that of the *soluta* physical equation, which is here measured by ABL. The latter is in the descending semicircle, since the moon at L is moving from apogee D to perigee F (and hence is descending toward the earth); so the *soluta* equation ABL is added to the circular sector DBL to give area DAL as the mean anomaly at L, with regard to the *soluta* inequality only. This calculation is exactly like that for a primary planet. Now, the evection ACL is also in its descending semicircle, since L is descending from

[57] *G. W.*, 7: 452: 24–33.
[58] *G. W.*, 7: 453: 19–28.
[59] Kepler explains the concept of a technical month in *G. W.*, 7: 449: 37–42, and uses the phrase frequently thereafter, for example 453: 29–33; 455: 12–14; 456: 11–12.
[60] *G. W.*, 7: 453: 21–28.

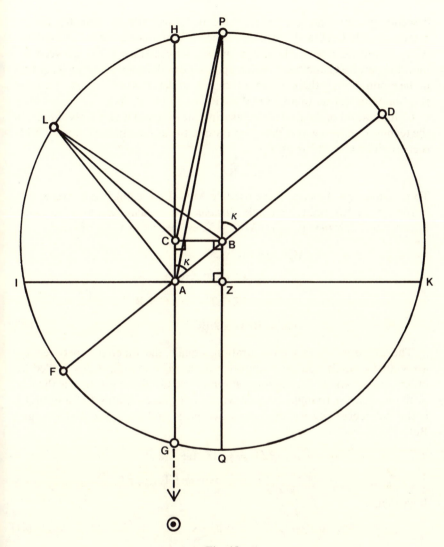

Fig. 45

monthly apogee H to monthly perigee G. Therefore ACL must be added to the previous area DAL to yield the mean anomaly at L with regard to the *soluta* inequality and the evection. The moon's position at L is called first-coequated (*primo coaequata*) with respect to area DAL, but second-coequated (*secundo coaequata*) with respect to area DAL + ACL.[61]

To clarify this further, let us take another case. If the moon were at P, the triangle measuring its separated inequality, ABP, would again be in the

[61] *G. W.*, 7: 454: 35–455: 14; 458: 21–32.

descending semicircle and hence would be added to area DBP to give the mean anomaly DAP with respect to which P is first-coequated. The evection, ACP, would here be in the ascending semicircle, since P lies between the monthly perigee G and the monthly apogee H, in the counterclockwise course of the moon. It must therefore be subtracted: so that P is the second-coequated position with respect to area DAP-ACP.

Computation of the triangle measuring the evection is relatively simple. In the technical month where the elongation of the apsidal line is κ, the monthly eccentricity AC or BZ is obviously

$$AC = BZ = e \cos \kappa \qquad (39)$$

To measure now the area of the triangle ALC (Figure 46), it is convenient to introduce the "associated triangle" (*triangulum socium*) BZL and the "withdrawn piece" (*particula exsors*) ABC.[62] These permit us to compute:

$$ACL = \tfrac{1}{2} \cdot AC \cdot LV$$
$$= \tfrac{1}{2} \cdot AC \cdot (LT - VT)$$
$$= \tfrac{1}{2} \cdot BZ \cdot LT - \tfrac{1}{2} \cdot AC \cdot CB$$
$$ACL = BZL - ABC \qquad (40)$$

The evection, in this configuration, equals the difference between the associated triangle and the withdrawn piece. (The modifications needed for other configurations are obvious: in semicircle QKP the evection is the sum of the associated triangle and the withdrawn piece; in arcs PH and GQ it is the difference between the withdrawn piece and the associated triangle.) But

$$BZL = \tfrac{1}{2} \cdot BZ \cdot LT$$
$$= \tfrac{1}{2} \cdot e \cos \kappa \cdot LT$$

If we define

$$\eta = \beta - \kappa \qquad (41)$$

(that is, η is the elongation of the moon from the syzygies as measured from the center of the orbit),

$$BZL = \tfrac{1}{2} \cdot e \cos \kappa \cdot \sin \eta \qquad (42)$$

On the other hand,

---

[62] *G. W.*, 7: 455: 17–25. The *particula exsors* is "withdrawn" from specific calculations in the sense that it is constant for a technical month, i.e., for given elongation of the apsidal line, and hence need not be calculated anew for different positions of the moon in its orbit. See Kepler to P. Crüger, 9 September 1624, letter No. 993, in *G. W.*, 18: 205.

$$ABC = \tfrac{1}{2} \cdot AC \cdot BC$$
$$= \tfrac{1}{2} \cdot e \cos \kappa \cdot e \sin \kappa$$
$$= \tfrac{1}{2} \cdot e^2 \cos \kappa \sin \kappa$$

which is constant for the technical month. Since we already have $ACL = BZL - ABC$ (40), then the triangle of evection is

$$ACL = \tfrac{1}{2} \cdot (e \cos \kappa \sin \eta - e^2 \cos \kappa \sin \kappa)$$

As before (above, p. 167) the factor $\tfrac{1}{2}$ disappears when we convert from area measure to the angular measure of mean anomaly. Thus in angular measure,

$$ACL = e \cos \kappa (\sin \eta - e \sin \kappa) \qquad (43)$$

The mean anomaly α for which L is the second-coequated position is therefore given, by (35) and (43), as

$$\alpha = DAL + ACL$$
$$\alpha = (\beta + e \sin \beta) + (e \cos \kappa [\sin \eta - e \sin \kappa]) \qquad (44)$$

Let us see how this evection behaves. When the regular apogee is at quadrature to the sun, $\kappa = 90°$ and $\cos \kappa = 0$. The monthly eccentricity $e \cos \kappa$, as given by (39), is therefore zero, as is the evection given by (43). On the other hand, when the regular apogee is in the syzygies, $\kappa = 0$, so that (41) and (39) reduce to

$$\eta = \beta$$
$$AC = e \cos 0° = e$$

The evection is then simply, from (43),

$$ACL = e \sin \beta \qquad (45)$$

That is, it equals the *soluta* physical equation, and will add an equal component to the equation of center (2;30° at its greatest, using Kepler's value $e = 0.04362$). Tycho, in fact, had found that the maximum evection equaled half of the maximum equation without evection: thus, in Kepler's terms, it equaled the maximum *soluta* physical equation. This meant that there was no need to renorm the area of the triangle ACL,[63] since at its greatest it coincided with the maximum *soluta* triangle ABL. Kepler could only attribute the equality here to simplicity and beauty,[64] which of course were quite sufficient reasons for God to have set things up this way.

---

[63] "... quod valde commodum accidit schematibus," *G. W.*, 7: 458: 1.
[64] *G. W.*, 7: 352: 1–2.

Kepler had reason for suspecting *a priori* that Ptolemy's value of about 2;40° for the maximum evection was correct, instead of Tycho's 2;30°. We shall take these up on p. 194.

We have been treating the orbit as an eccentric circle; it is, of course, an ellipse. We computed the triangle ACL of the evection as if L were on the circle, but a physically exact calculation would require that L be somewhat inside the circle. The error could not be very large, since the moon's orbit is constricted by only about $e^2 = .002$ of its radius; but in fact the error was very much smaller than this. The evection was maximal when the apsidal line was in the syzygies ($\kappa = 0$), when its triangle ACL coincided with the triangle ABL of the *soluta* physical equation. But in this case, as we saw in the planetary theory, areas calculated to the circle are entirely equivalent to areas calculated to the ellipse, because of the constant proportionality with which the ellipse divides ordinates perpendicular to the apsidal line. On Figure 46, imagine that the apsidal line coincides with GH, the line of syzygies. Then moving L inward from the circle to the ellipse, along the ordinate LV, reduces the area of the evection triangle by the constant proportion $\sqrt{(1 - e^2)}$ to 1. The proportions of these triangles among themselves are thus unchanged, as are their ratios to the similarly reduced areas measuring the *soluta* equation.

The configuration of greatest evection, therefore, involved no error at all in calculations which placed the moon on the circle. At the other extreme, when the apsides were at quadrature, orbital constriction would distort the proportions between the evection triangles and the *soluta* equations; but here the evection vanished, anyway. In intermediate locations there was some error; but it became proportionally smaller as the magnitude of the evection increased. Overall, then, Kepler was justified in calculating the evection on a circular orbit.

Physically, the evection moved the planet sometimes faster and sometimes slower than it would have moved were there no evection. This could only happen if the strength of the earth's rotating magnetic image changed in the presence of sunlight. Kepler believed that sunlight had just this effect. The strength of the image therefore varied in some way other than a simple inverse-distance relation. One may approach this question directly by means of the time-derivative of the eccentric anomaly, $d\beta/d\alpha$ (in simple Kepler motion $d\beta/d\alpha = 1/R$). Since Kepler thought the shape of the orbit to be unaffected by evection (that is, the evection was an entirely "physical" equation, with no "optical" part), and we have shown on pages 163–165 above that, on an ellipse, equal increments of β correspond to equal amounts of motion around the central body, this derivative expresses the moon's circumterrestrial velocity, and hence, in Kepler's physics, the strength of the circumterrestrial force.

Simpler, however, and at the same time more faithful to Kepler's own conceptions, is an evaluation of the mean anomaly, analogous to "Kepler's equation." In a technical month when the evection is maximal ($\kappa = 0$), equations (43) through (45) reduce to

$$\alpha = \beta + 2e \sin \beta \qquad (46)$$

The physical equation, which expresses the planet's deviation from uniform motion and which equals $e \sin \beta$ in simple Kepler motion, is precisely doubled. This was what Kepler meant in stating that it was "as if the motive image of the earth's body became precisely (*praecise*) twice stronger in close approach, weaker at a distance...."[65] Of course, that statement is not precise in a modern formulation of Kepler's theory. We are today inclined to derive

$$\frac{d\beta}{d\alpha} = \frac{1}{1 + 2e \cos \beta} \qquad (47)$$

from (46); to compare it to the corresponding expression without evection,

$$\frac{d\beta}{d\alpha} = \frac{1}{1 + e \cos \beta}; \qquad (48)$$

and to conclude that the force is not *praecise* twice stronger or weaker. It is not the "force equations" (47) and (48), but their reciprocals, in which the evection doubles the deviation from regular motion.

Such objections are misleading, however, for they impute to Kepler a conceptual exactness which his physics—in contrast to his astronomy—did not possess. We have repeatedly spoken of Keplerian forces as producing a velocity proportional to themselves, in order to distinguish them from Newtonian forces which produce acceleration. As we have seen, however, Kepler conducted his exact analysis in terms of delays, and not of velocities. The word force, *vis*, did not yet have the clarity of an accepted technical term. Kepler was violating no conventional usage, and perhaps was being meticulously precise, in asserting that the irregularity in a force had doubled when the irregularity in the delays doubled, as in (46).

Closely related to the evection or recurrent monthly inequality of longitude was the recurrent monthly inequality of latitude. The separated or *soluta* inequality of latitude has already been described as a constant 5° inclination of the lunar orbit to the ecliptic, caused by an axis in the moon's body. The recurrent monthly inequality of latitude varied this inclination as 5° ± 0;18° cos ω, where ω is the angle between the ascending node and the nearest syzygy. The effect described by this changing inclination was simply that the moon's "inequality" in latitude, that is, its motion away from or toward the mean ecliptic, was greater when the nodes were in the syzygies.

Kepler's physical account of the recurrent inequality of latitude was simple but not very precise. Just as the moon's motion in longitude was easier and swifter when it was traveling parallel to the circle of illumination (represented by IK in Figure 46), so also was its motion in latitude. The sphere of sunlight, after all, was extended in latitude as well as longitude. Most of the actual motion in latitude takes place near the nodes, since at the limits the orbit is

---

[65] *G. W.,* 7: 351: 41–43.

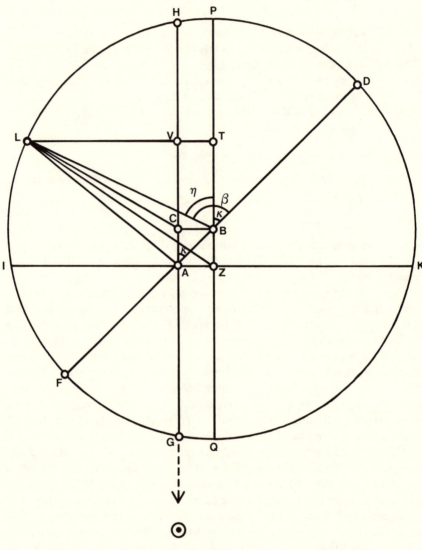

Fig. 46

parallel to the ecliptic. It was therefore when the nodes were near syzygy and
the moon near the nodes that motion in latitude was both appreciable in
magnitude and further strengthened by the sphere of sunlight—to which the
moon was then moving nearly parallel. On these occasions, one imagined the
inclination of the orbit to be increased slightly, to explain the accelerated
motion of the moon toward or from the ecliptic.

The most notable feature of this theory, qualitative as it is, is the neatness
with which it matches the account of the recurrent monthly inequality of

longitude. In a lunar theory adapted from the theory of primary planets, the moon's separated inequalities of longitude and latitude had been attributed to axes in the lunar body, axes which were directed toward the earth at the mean distances and the limits, respectively; and which were perpendicular to the radius from the earth at the apsides and the nodes, respectively. The theory of the moon had now revealed new, recurrent monthly anomalies which vanished when the syzygies coincided with the respective places where the axes pointed at the earth; and which reached their maximum when the syzygies coincided with the respective places where the axes were perpendicular to the radius from the earth. Surely Kepler took this as some small confirmation that he was on the right track. He was, of course, mistaken.

Of the lunar inequalities in longitude, we have thus far considered the separated or *soluta* inequality (responsible for the e sin β term in [44]) and the evection or recurrent monthly inequality (responsible for the e cos κ[sin η − e sin κ] term in [44]). Both of these were sometimes additive, sometimes subtractive, and hence cancelled out over long periods of time. We must now consider the third lunar inequality, the variation or "perpetual monthly inequality," as it was termed by Tycho and Kepler respectively. Kepler's account of the variation was much more interesting than his account of the evection. The perpetual monthly inequality, or variation, was always accelerative in Kepler's physics: it did not cancel out over time. Mathematically, of course, one could detach the mean cumulative effect of the variation and combine it with the mean motion, leaving a periodic remainder which did average to zero. Kepler did so, himself, for computational purposes. This should not obscure the fact that physically the variation was a fluctuating but always accelerative phenomenon.

The physical significance of the perpetual monthly inequality lay, like that of the recurrent monthly inequality, in the strengthening of the earth's rotating magnetic image by sunlight. This strengthening took place around the syzygies, where the moon was moving parallel to spheres of light centered on the sun. The effect of the strengthening was complicated by the changing distance between the earth and moon, as we have seen in the discussion of the evection. The perpetual monthly inequality, as its name implied, comprised only the mean acceleration from sunlight at syzygy, that is, the acceleration considered without regard to the changing distance from the earth.

We have seen that in Kepler's theory, the amount by which sunlight could strengthen the terrestrial image depended upon the angle at which they were "applied" to one another. In Figure 47, T is the earth, L the moon, S the sun; the circle LC is not precisely the lunar orbit but rather a circle centered on the earth, and representing the earth's rotating image at the present distance TL of the moon. The "line" LA is actually part of the circumference of a very large sphere of light centered on the distant sun. The angle between the spheres of sunlight and of virtue is thus the angle between the circle LC and the "line" LA. Since TA is perpendicular to the circumference LA, this angle is the complement of angle TLA; hence, assuming LS and TS' parallel, it is simply

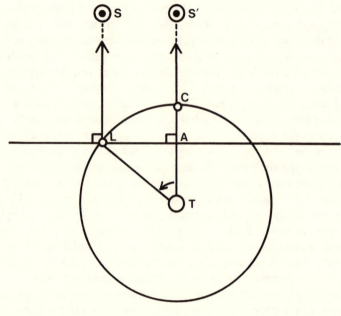

Fig. 47

angle LTS′, the elongation of the moon from the sun. We have been calling
this angle η.

It took Kepler a long time to work out the theory connecting the observed
variation with the elongation η. In a letter to Maestlin written in the spring
of 1620, he described the development of this theory and the reasoning behind
it.[66] He had begun, characteristically, by deriving from archetypal considerations the cumulative amount by which the moon's mean motion seemed to
be accelerated by the sun. There was no purely observational way to distinguish the component of the mean lunar motion which was due to the accelerative effect of sunlight. For Kepler, however, observation was only one way,
though the most reliable, of gaining insight into the structure of the universe.
Careful study of proportions could also reveal details of the plan by which
the universe had been created. Indeed, it would be no exaggeration to say that
this was Kepler's preferred method, and the goal of all his work—provided
that we remember always the painstaking care with which he used observations to evaluate, and when necessary to reject, his apprehensions of the
divine plan.

Kepler had found the relation he needed to estimate the solar contribution
to the moon's mean motion in the ratio of the month to the year. It seemed
to him that the strengths of the solar and terrestrial images, and the densities

---

[66] *G. W.*, 18: 17–20.

of the terrestrial and lunar bodies, must have been established according to a compelling archetype. Judging from the way things had turned out, the most likely pattern that was *nearly* realized was that a (sidereal) year should contain precisely twelve mean synodic months, each of precisely thirty days. In actuality, the moon moved a bit faster: a sidereal year included twelve synodic lunar revolutions and 132;45° of lunar motion besides.[67] The extra motion, Kepler thought, must surely derive from the strengthening of the terrestrial image by sunlight. During the amount of time when the earth's unassisted motive image would carry the moon through twelve complete revolutions, sunlight strengthened that image sufficiently to carry the moon another 132;45°. This calculation yielded the mean effect of sunlight on the moon. Kepler continued by explaining to Maestlin his reasoning about the inequalities in the effect of sunlight.

The extra velocity at any given time, according to Kepler's principles, should be proportional to "the apparent width, to the moon" (*apparentem ipsi Lunae latitudinem*) of the earth's circle of illumination. Kepler did not claim, he assured his teacher, that the appearance was itself a physical force, but only that "there is some motive cause, which uses this variation of the appearance as a law and rule, or a measure, of its motion."[68] (Kepler, as always, was willing to use a mathematical theory whose parameters were physically meaningful, even if he could not explain the mechanism through which the theory operated.) His first thought had been to distribute the extra 132;45° per year of motion according to the sines of the moon's elongation. The velocity which was responsible for this excess motion was therefore proportional to the cosine of the elongation. Since the apparent width, from any given point in the moon's orbit, of the circle of illumination on the earth was proportional to the cosine of the elongation (Figure 48), this simple hypothesis satisfied the requirements of his physics.

Calculations from this theory differed somewhat from Tycho's lunar observations, by as much as nine minutes of arc. Kepler had nonetheless preferred his theory to that of Tycho, which had no physical basis. In early 1620, when he recast his theory of the evection in preparation for the *Epitome*, he realized that the evection, which he was now treating as a purely physical phenomenon, could arise from the same source as the variation; and he therefore began trying to reconcile his partially-successful theory of the variation with the Tychonic results.

His theory did not move the moon fast enough in the syzygies, and therefore moved it too fast in the quadratures. Yet the cosines of the elongation, on which he based his theory, measured the apparent width of the earth's circle of illumination. He began to think more about his circle of illumination. Just as the earth moved the moon in two ways, because of its diurnal rotation and because of its illumination, so the moon was moved doubly, because it was a

---

[67] *G. W.*, 18: 17; see also *G. W.*, 7: 464: 16–24.
[68] *G. W.*, 18: 17.

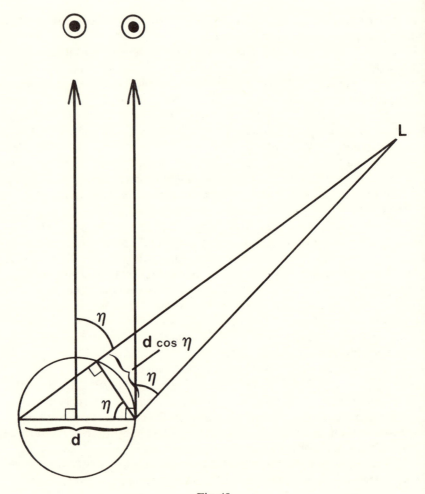

Fig. 48

solid body affected by the image of the earth's rotation and because it was itself illuminated by the sun. Let metaphysicians dispute about the possibility and manner of this, he avowed to Maestlin; he himself was compelled to investigate the measures of the motions he found in nature.[69] The motion of the moon indicated that its own illumination was an effective cause. The two circles of illumination on the earth and on the moon were always parallel, and their appearances, each as seen from the other body, were in phase. If this reasoning was correct, then the effect of the variation would be proportional not to the cosine of the moon's elongation from the sun, but to the squared cosine of the elongation. This adjustment brought his physically-based theory into line with Tycho's geometric theory.

---

[69] *G. W.*, 18: 18–19.

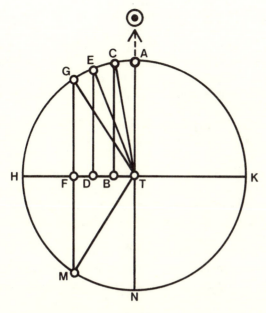

Fig. 49

The extra force imparted to the earth's image by the sunlight was therefore given by $\cos^2 \eta$. In the *Epitome* Kepler did not explain this relationship, which he had so recently arrived at, but simply stated it:

> the light of the sun aids the motive image of the earth, in the squared proportion of the cosines of the angles by which the image of the light of the sun, spread as a luminous spherical surface around the sun, and the spherical image of the body of the earth, spread around the earth, are mutually applied.[70]

The increment of force measured by $\cos^2 \eta$ produced in the planet, like all of Kepler's forces, a proportionate amount of added velocity, not of acceleration. Nevertheless a lunar theory was intended to predict position, rather than velocity, so Kepler had essentially an integration to perform. What was the cumulative effect on the planet's position of continual boosts in its velocity, each proportional to the squared cosine of the elongation η?

Kepler divided a circle (Figure 49) into small arcs, here shown as AC, CE, EG. In AC the additional velocity is then proportional to $\cos^2(ATA) = AT^2$; in CE to $\cos^2(ATC) = CB^2$; in EG to $\cos^2(ATE) = ED^2$; and so on. Now Kepler had already shown, in the context of his libration theory, that in the limit of infinitely small division, the sections intercepted on the diameter HK

---

[70] *G. W.*, 7: 461: 42–462: 2. Kepler claimed here, in Book Six, to have posited this relation in Book Four. This can only refer to his speculation mentioned earlier, in *G. W.*, 7: 323: 17–20, that if the variation only accelerated the moon, and never slowed it, it would have to follow the doubled or squared (*dupla*) proportion of that which would result simply from the angle of application.

were proportional to the perpendiculars intercepting them: that is, that TB, BD, DF, etc., were proportional to AT, CB, ED, etc. The pseudo-rectangles ATBC, CBDE, EDFG, etc., therefore were wide in proportion to their height, so that their areas were proportional to the square of their height. The areas were therefore proportional to the additional motion acquired: in moving from A to G the moon acquired additional forward motion, from sunlight, proportional to area ATFG.[71]

The simplest way to regard an area such as ATFG was to decompose it into a sector such as AGT, which increases uniformly with the elongation η, and a triangle such as TGF. Since the variation is physically and mathematically symmetric with respect to both the line of syzygies ATN and the line of quadratures HTK, the analysis of one quadrant suffices for the whole circle. At H, then, the added motion is proportional to the (quarter-circle) sector AHT, the triangle having vanished. At M, in the second quadrant, it is proportional to the same sector AHT plus the small area HMF; one expresses this most advantageously as sector AMT minus triangle MTF. The variation thus appears as the sum of, first, an element always positive and proportional to the elongation, and therefore in the long run proportional to time; and second, a correction proportional to the triangles GTF, etc., and hence to $\sin \eta \cdot \cos \eta = \frac{1}{2} \sin 2\eta$.

To determine the magnitude of the variation it sufficed in principle to know either of the two elements. The former could not be calculated from observation, since it was inextricably mixed with the mean motion. The latter, which was the actual inequality, had been found by Tycho to amount to about 0;40,30° in the octants, where it was greatest. This of course implied an equation amounting to 0;40;30° · sin 2η.

As explained above, Kepler thought that the extra motion (due to sunlight) in a sidereal year was probably equal to the 132;45° which the moon traveled in excess of its archetypal twelve synodic revolutions. This cumulative excess of 132;45° per year must therefore be proportional to the regularly increasing sector of the circle in Figure 49. (That sector represents the cumulative effect of the variation.) A sidereal year consisted of 12 synodic months of 360° each, plus 132;45°, or a total of 12.369 synodic months. He therefore computed the extra motion in a synodic month as $(132;45°/12.369) = 10;44°$, and hence the extra motion in one quarter of a synodic month as 2;41°. If the area of a quarter circle, which was 0.7853981634 as accurately as Kepler knew it, was equivalent to 2;41°, then the maximum equation of the variation, being equivalent to a triangle of area $\frac{1}{2} \cdot \sin 45° \cdot \cos 45° = .25$, must amount to

---

[71] It is clear from the figure that this geometrical result is equivalent to the standard integration formula

$$\int_0^\eta \cos^2 \eta \, d\eta = \frac{1}{2}\eta + \frac{1}{2}\sin \eta \cos \eta$$

where, with the moon at G, $\frac{1}{2}\eta$ is the area of sector AGT and $\frac{1}{2}\sin \eta \cos \eta$ is the area of triangle TGF.

$$\frac{.25}{.7853981634} \cdot 2;41° = 0;51°$$

Tycho's observations of this maximum equation had averaged to about $40\frac{1}{2}$ minutes instead of 51, as noted above, so there was some discrepancy.[72] Observe that in this theory, where the physical variation comprises both the equation, properly speaking, and the constant boost which merges with the mean motion, it is greater in magnitude than the evection. Sunlight always strengthened the earth's motive image in the syzygies. The strengthening was much less when apogee fell in the syzygy, but the net effect was always one of faster motion in the syzygies than would have occurred without the aid from sunlight.

Passages such as the one discussed above, where archetypal relations formed the basis for elaborate physical calculations, are extremely informative about Kepler's attitude toward those strange relations, which he was so fond of pursuing. Archetypes (by which Kepler usually seems to have meant ratios found in nature which were expressible in small whole numbers) are elusive to the modern mind. They strike us as unscientific, even "mystical" (a much abused word). We must remember that Kepler's universe had been created; and created with a definite mathematical plan which was known to the Creator, and not inaccessible to man. For him this was a confirmed fact, supported by reason and observation. It was by no means an "irrational" belief, still less a mystical inspiration, but rather an undoubted element of European culture. Kepler's archetypes often were not observed in the phenomena with any precision. This did not bother him, since he assumed that the archetypal ratios held among fundamental parameters which could not be observed directly. Some of these parameters were physical: the strengths of the rotating magnetic images, or the volumes and masses of planetary bodies. The phenomena could be deduced from the appropriate archetypes only when all of them were known, together with the physics by which they interacted to produce observable phenomena. I think this was a rather sophisticated attitude, not at all unscientific and very different from the number-mysticism which is sometimes carelessly attributed to Kepler.

Continuing his analysis of the peculiarities found in lunar theory, Kepler argued that the moon's accelerated motion due to sunlight also accounted for the comparatively rapid movement of the lunar apsides and nodes.[73] Recall that in the physical account of simple Kepler motion, the fibers of libration were deflected slightly from their normal position perpendicular to the apsidal line. This deflection happened at such a rate that the fibers pointed directly at the sun when the planet was halfway from aphelion to perihelion: at E in Figure 50a. Thereafter the deflection diminished, so that they were again perpendicular to the solar radius at G. Therefore G was perihelion, for the

---

[72] G. W., 7: 464: 8–20.
[73] G. W., 7: 352: 42–353: 27.

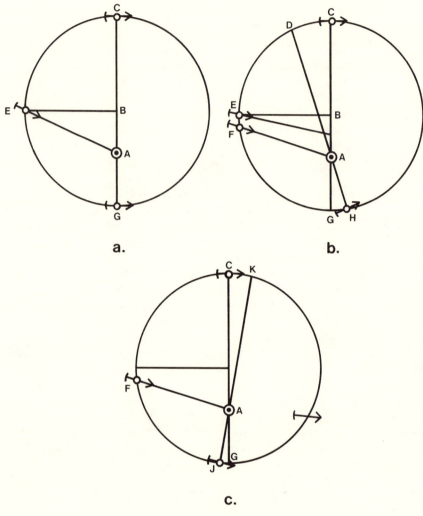

a.   b.

c.

Fig. 50

fibers in this position were neither attracted nor repelled by the sun. Now
consider the moon, which moved faster than it should, because of the aid from
sunlight. The tugging which deflected the fibers of libration was not affected
by sunlight. The moon reached E (in Figure 50b) quicker than it would
otherwise, and hence before the fibers had time to deflect far enough to point
at the sun. They finally pointed at the sun only at some lower place, say F.
Therefore, implied Kepler, they would be restored perpendicular to the sun's
radius only at some point H beyond G. Line HAD would be the apsidal line
then, and the apsides would have moved by angle HAG (empirically about
1;32°) in half an anomalistic month.

This explanation simply does not work. One would like an explanation of why the strengthened terrestrial image, which does attract and repel the moon in proportion to its augmented vigor, *via* the fibers, does not also deflect the fibers in that proportion; but that is a relatively small deficiency. More important is Kepler's failure to think through the physics of the second and later quadrants. From F, where we left the moon last paragraph, it would have been carried swiftly toward G while its fibers were slowly deflected back counterclockwise. Evidently (Figure 50c) they would be perpendicular to the solar radius at some point J *before* reaching G. This point J would be the new perihelion, and line JAK the new apsidal line. It has moved in the wrong direction. Similar considerations show that the apsidal line would continue to retrogress in the third and fourth quadrants.

The failure of this small theory, to which Kepler devoted only a couple of pages, does not invalidate those on which it is based. The simple assumption that the deflection of the fibers, like their attraction, followed the strength of the terrestrial *vis prensandi* would suffice to leave the apsides unmoved, awaiting a workable theory of their progression.

The rapid motion of the lunar nodes, compared to those of the primary planets, did not present such severe difficulties. Kepler's theory for the primary planets (above, Figure 42) had been that the solar image exerted a pressure on the fibers of latitude, and that this pressure was very slightly greater on the side of the planet toward the sun. The pressure differential slowly rotated the fibers, and hence the nodes, *in antecedentia*. Kepler now suggested that the moon's fibers of latitude, since they were "not guarded against" the strengthened force from sunlight, responded much faster to the terrestrial image than did the fibers of the primary planets to the solar image.[74] As in the archetypal computations of the variation discussed above, an extraneous physical force disturbed the harmony which would have prevailed in its absence.

Before leaving Kepler's lunar theory we must consider his attitude toward what is now known as the annual equation of the moon. Although Kepler was one of the first, along with Tycho, to identify this equation, he thought it to be a part of the equation of time rather than of the lunar motion.[75] Accordingly we must give a brief account of the equation of time.

Astronomical time in the early seventeenth century was still measured by the diurnal passage of the sun, the "apparent day." An apparent day is simply the interval of time between two successive meridian transits of the sun. A meridian transit occurs when the sun crosses the meridian, the arc through the zenith connecting the north and south celestial poles. It is of some importance to note that the meridian is what principally moves, as the earth rotates, catching up with the much slower motion of the sun in the ecliptic at noon each day. Since the sun is indeed moving in the ecliptic, an apparent day is

---

[74] G. W., 7: 353: 27–34.
[75] There is more to the story than this, but the complications would distract us from our subject. See Anschütz, "Über die Entdeckung der Variation," cited above.

slightly longer than the time required for the earth to rotate through 360°. (The most convenient viewpoint for the following is quasi-geocentric. The sun moves in the ecliptic, for narrative simplicity; but the diurnal rotation pertains to the earth, because this physical truth is what we will be examining.) After turning through 360°, the meridian must catch up with the sun, by turning through the angle of the sun's daily motion in right ascension, which is on average slightly less than a degree. Because of the periodic changes in the apparent motion of the sun, the length of an apparent day varies. Therefore, to obtain the true time interval between two astronomical observations one must adjust for this irregularity.

The equation of time measures the irregularity in the length of the apparent day. It has two components. The first is due to the fact that the sun moves in the ecliptic, while the earth's rotation is along the equator. Where the ecliptic and equator are parallel, at the solstices, the sun moves directly away from the meridian as the latter sweeps across the heavens. Where the two circles cross and make their greatest angle, at the equinoxes, the sun moves away from the meridian at an angle, and so is more quickly overtaken. The first component of the equation thus has a period of half a tropical year, the period between solstices.

The second component of the equation of time follows from the non-uniform motion of the sun in the ecliptic itself. The sun's apparent motion is slower when it is more distant, in the squared ratio of the distances. Therefore, after allowing for the first component, an apparent day will be longer when the earth is at perihelion, and the sun appears to move further during the course of a terrestrial rotation. This second component has a period of an anomalistic year.

Both components are easily calculated from elementary spherical astronomy and solar theory. They find their principal use in lunar theory: the moon moves an average of half a degree an hour, so that a mistake of ten minutes in the time between two observations will produce an error in position perceptible to the naked eye.

The events which led Kepler to postulate a third component of the equation of time are described only very briefly in the *Epitome*.[76] Tycho had found that his lunar theory agreed better with observation if he neglected to correct for the second, annual component of the equation of time. Kepler of course realized, as early as 1600,[77] that this was not an acceptable way of proceeding. The second component followed inescapably from solar theory, and could not be discarded without pretending that the sun moved uniformly in the ecliptic. However, his own lunar theory suggested that sunlight accelerated the moon's motion by increasing the strength of the earth's magnetic image. Perhaps it also accelerated the earth's rotation. If the earth rotated faster because of sunlight, then the intensity of the sunlight should affect the speed of rotation.

---

[76] *G. W.*, 7: 184: 3–5, 409: 19–24.
[77] Letter to Archduke Ferdinand of Austria, #166 in *G. W.*, 14: 125: 220–234.

Specifically, the earth should rotate faster at perihelion, so that the apparent day would consequently be shorter; and slower at aphelion, so that the apparent day would last longer. This irregularity would be precisely opposite in phase to the second component of the equation of time, and would explain why Tycho's lunar theory had fit observations better when that component was neglected.

This "physical equation of time," which resulted from sunlight varying the speed of the earth's rotation, raised a number of questions. If the earth spun faster on its axis at perihelion, would it not then move the moon proportionally faster? It seemed that the two effects, that upon lunar motion and that upon apparent time, should cancel one another observationally. The English prodigy Jeremiah Horrocks, perhaps Kepler's most profound student in the next generation, raised this question in a letter of 1637.[78] Some of Kepler's fragmentary notes seem to indicate that he anticipated the objection. His very tentative solution was to suppose that it must be simply the earth's motion, and not the speed of that motion, which set the moon spinning around it.[79]

The *Epitome* contains no observational data on the magnitude of the physical equation of time. Instead there is a purely theoretical computation from physics. The pattern of variation was simply and elegantly calculated. The sun's assistance to the terrestrial rotation was surely attenuated just as sunlight in general was attenuated in longitude, namely in inverse proportion to the distance between earth and sun. This pattern was, of course, the same which governed the earth's annual revolution about the sun. For the circumsolar motion as well as the solar supplement to the earth's own rotation, the variation followed the inverse proportion of the distance separating earth from sun. The cumulative effect of the varying solar boost to the rotation was the physical equation of time. The cumulative effect of variation in the earth's circumsolar motion was the physical equation of center. The former equation, then, was proportional to the latter. To calculate the physical equation of time Kepler therefore used the device invented to calculate a planet's physical equation of center: the area law![80]

Kepler had surprisingly little to say about the application of the area law to an irregularity of the diurnal rotation. The analogy he used, however, was a complex one and requires careful examination. The familiar terms in the analogy were those of the area law: the delay per unit of circumsolar motion was proportional to the distance, and the accumulated delay over some interval of circumsolar motion was proportional to the area swept out by that motion. The new element was the supplement to the earth's rotation, and this supplement varied inversely as the earth's distance from the sun. That is, the delay or time elapsed while the solar boost increased the earth's rotation by some unit was proportional to distance. Therefore, by Kepler's analogy, the

---

[78] Horrocks to G. Crabrius, August 12, 1637, quoted in Frisch, *O. O.*, 6: 562.

[79] Frisch, *O. O.*, 6: 600.

[80] *G. W.*, 7: 408: 39–409: 17.

accumulated extra time over the mean time[81] required to complete some amount of supplementary rotation was proportional to the size of the sector.

We must now ask what determined the magnitude of the sector. For the original area law (Figure 39), the sector intercepted an arc PG determined by the eccentric anomaly, which, as we have seen, was a measure of the total circumsolar motion accomplished. To maintain the analogy, the sector measuring the time required for supplementary rotations must intercept an arc PG determined by the amount of *supplementary* rotation experienced by the earth.

This will become clearer if we supply a few numbers. We adopt Kepler's value for the total quantity of the additional rotation due to sunlight. The archetype for unassisted terrestrial rotation, he felt sure, was 360 days in a year. The $5\frac{1}{4}$ rotations beyond these which the earth in fact experienced were therefore due to sunlight. Now, let us for a moment forget that Figure 39 represents the earth's orbit. Let the circumference of the whole circle PGR represent the 1890° of supplementary rotation which are caused by sunlight in a year ($5\frac{1}{4} \cdot 360° = 1890°$). If this extra rotation had taken place at a uniform rate, then the time elapsed during any part of it would have been a simple fraction of the year. A quarter of the extra rotation, $472\frac{1}{2}°$, would be represented by a quarter of the circumference, arc PD. At a uniform rate, this would of course take place in a quarter of the whole year, represented by area PBD, a quarter of the circle. (In fact, if the sun were at the center B, the supplement would be uniform and a quarter of it would have taken place after a quarter of the year.) But the rotation took longer during those parts of the orbit more distant from the sun at A: so the actual time needed to turn an extra $472\frac{1}{2}°$ was instead proportional to area PAD.

At this point the analogy gets complicated. We were just now insisting that the circle PDR represent not the orbit, but the 1890° of sun-induced rotation which the earth experienced in a year. The area PAD of the sector represents the mean anomaly not so much when the earth is at the quadrant of its orbit, but rather when a quarter of this supplementary rotation has occurred. How do we know where the earth will be after a quarter of its supplemental 1890°? The supplement follows the same proportion as the circumsolar component of the earth's annual motion (namely, the inverse of the distance). As Kepler showed, though, the measure of circumsolar motion is simply the eccentric anomaly. It is entirely proper, then, to let circle PDR represent not only the supplementary rotation of the earth's axis, but also the eccentric anomaly of the earth's annual revolution.

The maximum of the equation, then, is given by the greatest area of triangle ABD (still in Figure 39), where the whole circle is $5\frac{1}{4}$ days. Since Kepler's value for the eccentricity of the earth's orbit was .018, we have for the maxi-

---

[81] On the other hand, after perihelion the earth's extra rotational motion had been consuming less time than usual. In the ascending half of the orbit, then, the equation measured the accumulated deficiency of time needed to perform these rotations, below the mean.

mum physical equation of time τ:

$$\tau_{max} = 5\tfrac{1}{4}\,\text{days} \cdot \frac{\text{area ABD}}{\text{area of circle}}$$

$$= 5\tfrac{1}{4}\,\text{days} \cdot \frac{.009}{3.1415926536}$$

$$= 21 \text{ m. } 40 \text{ s.}$$

This is Kepler's result. Accurate computation with the above numbers, which are from the text, gives 21 m. 39 s., but Kepler may not have used the precise value $5\tfrac{1}{4}$ for the excess of a year over 360 days. The equation is additive (i.e., to apparent time, to get mean time) in the half-year from aphelion to perihelion.

These were the elements of Kepler's lunar theory. (The physical equation of time, of course, was not a part of lunar theory; but it was only lunar calculations which really required the application of this equation.) As a whole, the theory exhibits a notable simplicity of conception, together with the highest level of mathematical sophistication to be found in Kepler's astronomical work. The elegant parallels between the recurrent inequalities of longitude and latitude, the surprising ease with which physical calculation intertwines with archetypal speculation, and the intricacy of mathematical development all mark Kepler's lunar theory as one of the loveliest of his scientific achievements.

Chapter 5

Kepler and the Development of Modern Science

I have concentrated exclusively, thus far, upon the internal development of Kepler's physical astronomy. In doing so I have tried to broaden the perspective of those who study only the parts of his work which were "correct," and equally of those who emphasize only that a great scientist had theories which seem wrong, or even (to abuse the word) "mystical." One gains nothing in evaluating Kepler's science according to anachronistic standards; but on the other hand, his arguments were rational, and we can follow his reason if we attend it carefully. Now that we have done so for an important part of Kepler's work, let us step back and try to re-evaluate his peculiar role in the larger drama known as the scientific revolution.

In most accounts of seventeenth-century science, Kepler stands somewhat apart. Although his work is a preëminent instance of the extraction of general laws from careful study of observations—a process which is obviously central to natural science—he does not fit, somehow, into the main plot line of the development of modern science. He completed the structure of ancient astronomy at just the time when the historian's spotlight shifts to the birth of modern physics in the contemporary work of Galileo. The importance we attach today to Kepler, in singling out part of his work as "the laws of planetary motion," is largely due to the value which accrues to those laws in a science completely unknown to Kepler, classical or Newtonian physics. They arouse our wonder not so much because they are accurate (for planets do not move exactly in accord with those laws), but because they are exactly the formulations one obtains in important idealizations of Newtonian dynamics. The fact that Kepler had already published them, more than half a century before Newton's derivations, was obviously of critical importance for the acceptance of the *Principia*. Equally obvious is that Newton's derivations account for the later enshrinement of "Kepler's laws" among the great achievements of science. The vicarious hypothesis could never have attained the same stature, because its theoretical interest was limited to ancient astronomy.

It can be said of Kepler, as of very few great scientists, that what he accomplished would never have been done had he himself not done it. It is in the nature of science that one seldom can say this about important achievements. If Isaac Newton had died in the plague year the development of mechanics would have been slower, but the mathematicians of the eighteenth century would eventually have obtained all his important results. If William Harvey had been content to treat his patients, someone else would have written (perhaps not so persuasively) on the circulation of the blood. Historians are wary of counterfactual assertions like these, but the compelling logic of a developing science makes it hard to doubt the outcome in such cases. With Kepler it is different. The discovery from examination of naked-eye observational reports that planets move on ellipses, and according to the area law, is so exceedingly improbable—and Kepler's manner of arriving at it was so decidedly personal—that it lies well outside the course of any inevitable development. The eventual derivation of such motion on the basis of classical physics was certain; its empirical discovery without the guidance of classical physics was really quite extraordinary.

In other words, the reason that Kepler's greatest discoveries surprise us, and that his place in history is so hard to pin down, is that his work transcends the science of ancient astronomy without arriving at classical physics. Much of what he did, of course, was entirely consistent with the characteristic problems, methods, and goals of ancient astronomy. We can easily understand refinements such as the use of the true sun rather than the mean sun, and the dismissal of the intricacies of latitude theory in favor of motion in an inclined plane, as the completion of Copernicus's work. They were an essential part of the conceptual simplification which the heliocentric transformation made possible. Careful analysis of technical details of the models in *De revolutionibus* forced Kepler—and would have forced others—to conclude that if one was going to be a Copernican one surely had to adopt these modifications. Even at a more sophisticated level, Kepler's work on the longitudes of Mars in Part Two of the *Astronomia nova* (which preceded his physical analysis, and which he described as "imitation of the ancients") was but an elaboration of the methods of his predecessors. It was in the third part of that book that he undertook a new astronomy; and had he not done so he could never have discovered "Kepler motion."

The assertion that planetary orbits were ellipses resembled in no way the merely technical improvements to Copernicus's models. Not only was it qualitatively different from any earlier supposition about the orbits; it answered a question that had not been raised, except as an afterthought, by any mathematical astronomer since the infancy of that science. What was the shape of the orbit, the actual path traversed by the moving planet? To earlier astronomers the fundamental concepts were geometrical, and the path as a whole was a composite phenomenon best understood by analyzing it into its components. The physical motion of the planet was thought less intelligible than the geometry astronomers introduced to explain it. Ironically, the simplicity they achieved by this analysis had distracted even Copernicus from recognizing that finally in his system each planet had an individual (and potentially a simple) orbit. By eliminating the Ptolemaic epicycles he had cleared the way for this realization; but in his dogged adherence to the standard of uniform rotation he spurned it. Yet even if he had retained the equant, and with it a perfectly circular path for the planet, he conceived his task to be one of reducing the phenomena to geometry.

Kepler's search went deeper. He was not satisfied with a geometrical theory until he had further reduced it to physics. This was why he asked novel questions about the source, direction, and intensity of the forces moving the planet. These questions were what enabled him, after taking ancient astronomy to its highest level of accuracy, to discover and to recognize results which transcended that science. They were questions which strained Kepler's geometrical powers to their limits, for the ellipse and the distance (or area) law are discoveries whose investigation naturally pertains to analysis, the language of classical physics, and not to geometry, the language of ancient astronomy.

It is these aspects of Kepler's work—his appreciation of planetary motion as a physical process to be understood in physical terms, and his posing of the novel questions raised by his new outlook—to which we must turn if we are to understand how he was able to discover his laws of planetary motion by studying Tycho's observations. In fact, his discovery was not simply an empirical one. Kepler was guided by no very sound theory in arriving at elliptical orbits and the distance law, but he was guided by the *right kind* of theory, a dynamic physics of forces between material bodies. Without it he could never have built his new astronomy.

I have pointed out from time to time that even Kepler's most tentative and abstract theories were of help to him. Time and time again he analyzed a problem into extremely general physical terms, either without any positive theory or with one having only the vaguest substance (a planetary "mind," for instance). The physical study of astronomical problems, even at so abstract a level, impelled him to ask a series of critical questions. From where could the motion originate? Where was it directed? How, roughly, did its components seem to vary? In what physically intelligible way might they *really* be varying, in order to appear like that? What exactly would be the outcome if they behaved intelligibly, in that way? We have seen Kepler follow lines of inquiry like this with many different degrees of detail and assurance, and with outcomes ranging from not at all helpful to wonderfully exact. At such times his investigations followed the great scheme of physical analysis and synthesis which Newton later stated so succinctly: from the phenomena to infer laws, and from these to deduce new phenomena.

It would be trite to attribute Kepler's success to so general a precept, but it would be unwise not to notice that he worked at the very time when the precept was becoming so persuasive. Kepler was the first astronomer whose analysis reached laws of nature that were physical rather than mathematical or philosophical. Only after he had reduced the observed phenomena to a theory of forces between material bodies could he see how to synthesize such new and complex relations as elliptical orbits or the distance law. The discursive account in the *Astronomia nova* reveals, in unusual detail, the development of Kepler's thought along this path. In other aspects of his work, notably in mathematical technique and in the use of observational data, the record shows an equally remarkable growth in sophistication, although I have not attempted to trace these paths to any depth.

I have tried to follow faithfully the development of Kepler's physical theory from its abstract, imprecise beginnings to the mathematically sophisticated theories from which he was finally able to deduce his most important astronomical conclusions. It sounds odd to say that Kepler 'deduced' his laws of planetary motion; he certainly did not in any straightforward sense discover his results by physical deduction. Yet there is a sense in which that is precisely how he discovered them, if by "discover" we mean to recognize that these really were possible solutions to important problems. Much more than this is involved in a discovery: someone must first notice the solution, by speculation, exploratory calculation, or other means; and someone must test the discovery sufficiently to give it credibility as the right solution. It is no small thing, however, to insist that Kepler's deduction of his astronomy from his physics elevated (for him) the hypotheses, from accurate but arbitrary suppositions to potential solutions of the great problem in the new astronomy: how did the planets really move?

Kepler was never content with an astronomical model he could not explain physically. After thinking of the idea that the Martian orbit was an ellipse, Kepler did *not* undertake calculations to prove that this was the correct shape. By then he already knew, as he said (see p. 129, above), that such calculations would be satisfactory. And satisfactory agreement, although it was all that astronomical calculations could provide, was not enough. Instead, he set out immediately to determine whether he could deduce an elliptical orbit from his physics. When he succeeded in that derivation he finally closed his long investigation of Mars.

It no longer seems quite so extraordinary, I hope, that the first astronomer who diverted his science toward its modern place within physics was also the one who was able to discover the relations on which Newton founded modern astronomy. These relations—"Kepler's laws"—could not be appreciated until physical questions were asked. I have tried to show in some detail that his physical investigations served an essential purpose for Kepler. They guided him in ordering and filtering the myriad possibilities which faced him when he denied the old principles of the uniformity, and then of the circularity, of planetary motion. His physics, although wrong, served him rather well; and the reason for this is not hard to see. It was not a philosophical physics, based on axioms and logic, but an empirical physics raised upon the firm foundation of knowledge built up by the Greek, Arab, and Latin students of ancient mathematical astronomy.

Bibliography

Books

Aiton, E. J. *The Vortex Theory of Planetary Motion*. New York: American Elsevier, 1972.

Beer, Arthur, and Beer, Peter. *Kepler: Four Hundred Years*. Volume 18 in *Vistas in Astronomy*. New York: Pergamon, 1975.

Caspar, Max. *Bibliographia Kepleriana*. Munich: Beck, 1936.

————. *Kepler, 1571–1630*. Translated by C. Doris Hellman. New York: Collier, 1959.

Copernicus, Nicolaus, *De revolutionibus orbium coelestium*. First published Nuremburg, 1543. Facsimile reprint New York: Johnson Reprint Company, 1965.

Delambre, J. B. J. *Historie de l'Astronomie Moderne*. 2 volumes. First published Paris, 1821. Reprint New York: Johnson Reprint Company, 1969.

Dreyer, J. L. E. *A History of Astronomy from Thales to Kepler*. New York: Dover, 1956.

————. *Tycho Brahe*. New York: Dover, 1963.

Euclid, *The Elements*. Translated by Thomas Heath. 3 volumes. New York: Dover, 1956.

Gilbert, William. *De magnete*. Translated by P. F. Mottelay. New York: Dover, 1958.

Heath, Thomas. *A History of Greek Mathematics*. 2 volumes. New York: Dover, 1981.

Herz, Norbert. *Geschichte der Bahnbestimmung von Planeten und Kometen. II Theil: Die Empirischen Methoden*. Leipzig: Teubner, 1894.

Kepler, Johannes. *Gesammelte Werke*. Edited by W. Van Dyck, M. Caspar, F. Hammer. Munich: Beck, 1937–.

————. *Neue Astronomie*. Translated by Max Caspar. Munich: R. Oldenbourg, 1929.

————. *Opera Omnia*. Edited by C. Frisch. Frankfurt: Heyder & Zimmer, 1858–91.

Koestler, Arthur. *The Sleepwalkers*. London: Hutchinson, 1959.

Koyré, Alexandre. *The Astronomical Revolution*. Translated by R. E. W. Maddison. Ithaca: Cornell University Press, 1973.

Neugebauer, O. *The Exact Sciences in Antiquity.* New York: Dover, 1969.

———. *A History of Ancient Mathematical Astronomy.* 3 volumes. New York: Springer-Verlag, 1975.

Newton, Isaac. *Philosophiae Naturalis Principia Mathematica.* First published London, 1687. Facsimile reprint London: W. Dawson & Sons, n.d.

Pedersen, Olaf. *A Survey of the Almagest.* N.P.: Odense University Press, 1974.

Ptolemy, Claudius. *The Almagest.* Translated by G. J. Toomer. New York: Springer-Verlag, 1984.

Small, Robert. *An Account of the Astronomical Discoveries of Kepler.* First published London, 1804. Reprint Madison: University of Wisconsin Press, 1963.

Swerdlow, Noel, and Neugebauer, O. *Mathematical Astronomy in Copernicus's De Revolutionibus.* 2 volumes. New York: Springer-Verlag, 1984.

Articles

Aiton, E. J. "Infinitesimals and the Area Law." In *Internationales Kepler-Symposium.* Edited by F. Krafft, K. Meyer, B. Stickler. Hildesheim: Gerstenberg, 1973. Pp. 285–305.

———. "Kepler's Second Law of Planetary Motion." *Isis* 60 (Spring 1969): 75–90.

Anschutz, C. "Ueber die Entdeckung der Variation und der jährlichen Gleichung des Mondes." *Zeitschrift für Mathematik und Physik, Historisch-Literarisch Abtheilung* 31 (1886): 161–171, 201–219; 32 (1887): 1–15.

Bialas, V. "Die Bedeutung des dritten Planetengesetzes für das Werk von Johannes Kepler." *Philosophia Naturalis* 13 (1971): 42–55.

Drake, Stillman. "Introduction." In Galileo Galilei, *Two New Sciences.* Madison: University of Wisconsin Press, 1974. Pp. ix–xxx.

Gingerich, Owen. "Kepler's Treatment of Redundant Observations." In *Internationales Kepler-Symposium.* Edited by F. Krafft, K. Meyer, B. Stickler. Hildesheim: Gerstenberg, 1973. Pp. 307–314.

———. "Kepler." In *Dictionary of Scientific Biography.* Edited by Charles C. Gillispie. New York: Scribner, 1970-. 7: 289–312.

———. "The Origins of Kepler's Third Law." In Arthur Beer and Peter Beer. *Kepler: Four Hundred Years.* New York: Pergamon, 1975. Pp. 595–601.

Goldstein, B. "The Arabic Version of Ptolemy's Planetary Hypotheses." *Transactions of the American Philosophical Society*, New Series, 57 (June 1967): 3–12.

Grafton, Anthony. "Michael Maestlin's Account of Copernican Planetary Theory." *Proceedings of the American Philosophical Society* 117 (December 1973): 523–550.

Hoyer, Ulrich. "Kepler's Celestial Mechanics." *Vistas in Astronomy* 23 (1979): 69–74.

———. "Über die Unvereinbarkeit der drei Keplerschen Gesetze mit der Aristotelischen Mechanik." *Centaurus* 20 (1976): 196–209.

Kennedy, E. S. "Late Medieval Planetary Theory." *Isis* 57 (1966): 365–378.

Kennedy, E. S., and Roberts, Victor. "The Planetary Theory of Ibn al-Shatir." *Isis* 50 (1959): 227–235.

Krafft, Fritz. "Johannes Keplers Beitrag zur Himmelsphysik." In *Internationales Kepler-Symposium.* Edited by F. Krafft, K. Meyer, B. Stickler. Hildesheim: Gerstenberg, 1973. Pp. 55–139.

Maestlin, Michael. "De dimensionibus orbium et sphaerarum coelestium." In Johannes Kepler, *Gesammelte Werke*. Edited by W. Van Dyck, M. Caspar, F. Hammer. Munich: Beck, 1937-. 1: 132–145.

Neugebauer, O. "On the Planetary Theory of Copernicus." *Vistas in Astronomy* 10 (1968): 89–103.

Price, Derek. "Contra-Copernicus: A Critical Re-estimation of the Mathematical Planetary Theory of Ptolemy, Copernicus, and Kepler." In *Critical Problems in the History of Science*. Edited by M. Claggett. Madison: University of Wisconsin Press, 1959. Pp. 197–218.

Riddell, R. C. "The Latitudes of Venus and Mercury in the Almagest." *Archive for History of Exact Sciences* 19 (1978): 95–111.

Roberts, Victor. "The Solar and Lunar Theory of Ibn Ash-Shatir." *Isis* 48 (1957): 428–432.

Russell, J. L. "Kepler's Laws of Planetary Motion: 1609–1666." *British Journal for the History of Science* 2 (1964): 1–24.

Swerdlow, Noel. "The Derivation and First Draft of Copernicus's Planetary Theory: A Translation of the *Commentariolus* with Commentary." *Proceedings of the American Philosophical Society* 117 (December 1973): 423–512.

———. "The Origin of Ptolemaic Planetary Theory." (Unpublished manuscript.)

———. "Pseudodoxia Copernicana." *Archives internationales d'histoire des sciences* 25 (1975): 109–158.

Thoren, Victor. "Tycho Brahe's Discovery of the Variation." *Centaurus* 12 (1967): 151–166.

Treder, H. -J. Die Dynamik der Kreisbewegungen der Himmels-körper und des Frien Falls bei Aristoteles, Copernicus, Kepler und Descartes. *Studia Copernicana* 14 (1975): 105–150.

———. "Kepler und die Begründung der Dynamik." *Die Sterne* 49 (1973): 44–48.

Wesley, Walter G. "The Accuracy of Tycho Brahe's Instruments." *Journal for the History of Astronomy* 9 (February 1978): 42–53.

Westman, Robert S. "Three Responses to the Copernican Theory: Johannes Praetorius, Tycho Brahe, and Michael Maestlin." In *The Copernican Achievement*. Edited by R. S. Westman. Berkeley: University of California Press, 1975. Pp. 285–345.

Whiteside. D. T. "Keplerian Planetary Eggs, Laid and Unlaid, 1600–1605." *Journal for the History of Astronomy* 5 (1974): 1–21.

———. "Newton's Early Thoughts on Planetary Motion: A Fresh Look," *British Journal for the History of Science* 2 (1964): 117–137.

Wilson, Curtis. "Kepler's Derivation of the Elliptical Path." *Isis* 59 (1968): 5–25.

———. "The Inner Planets and the Keplerian Revolution." *Centaurus* 17 (1972): 205–248.

Glossary

Acronychal observations: observations of a planet made at the moment when its second anomaly has vanished. In heliocentric theories, an acronychal observation is one made when the earth is directly on a line between the sun and the body being observed, so that the observed direction to the body is also the direction from the sun to that body.

Anomaly: (1) a periodic variation or irregularity in the motion of a heavenly body; also inequality. See first anomaly; second anomaly. (2) the position of a heavenly body in its orbit; conventionally expressed, in Kepler's time, as an angle measured from apogee or aphelion. See mean anomaly; eccentric anomaly; true anomaly; coequated anomaly; epicyclic anomaly.

Aphelion: the point of greatest distance from the sun on an eccentric orbit (in a heliocentric context).

Apogee: the point of greatest distance from the earth on an eccentric orbit (normally in a geocentric context). Kepler sometimes used this term for the point of a planet's greatest distance from the mean sun, since that is by definition its greatest distance from the mean position of the earth.

Apsidal line: the line connecting the apsides.

Apsides: aphelion and perihelion (or apogee and perigee), collectively. Singular form: apse.

Area law: (A modern term.) Kepler's discovery that the time required to traverse any arc of a planetary orbit is proportional to the area of the sector between the central body and that arc. See also distance law.

Bisected eccentricity: the Ptolemaic hypothesis wherein, in an eccentric orbit, the eccentricity of the orbital center was half of the eccentricity of the equant center.

Coequated anomaly: Kepler's usual term for the true anomaly, the angular distance of the planet from aphelion, as measured from the central body.

Concentric: a circle or sphere, in an astronomical model, in which the central body or point is exactly at the center. Opposed to eccentric.

Conjunction: the situation or moment in which a heavenly body lies at the same geocentric longitude as the sun, and hence is directly on a line from the earth through the sun. See opposition.

Delay: the amount of time required for a planet to traverse some small arc in its orbit. Kepler said that a planet's delays increased, rather than saying that its velocity decreased.

Distance law: (A modern term.) Kepler's theory that the speed with which a planet was carried around the sun varied in inverse proportion to its distance from the sun. Eventually he formulated the area law as a computationally convenient restatement of this theory.

Diurnal parallax: the tiny daily variation in the apparent location of a heavenly body (usually the moon) which is due to the rotation of the earth, carrying the observer with it.

Eccentric: a circle or sphere, in an astronomical model, whose central body is *not* exactly at the center. Such a situation results in an apparent irregularity of observed motion, even if the actual motion is regular.

Eccentric anomaly: normally, the distance of a heavenly body from aphelion or apogee, as measured by the angle at the center of the (eccentric) orbit, or by the arc length on the orbit itself.

Eccentricity: in an eccentric orbit, the distance between the central body and the center.

Ecliptic: the plane of the sun's orbit, in geocentric theory, or of the earth's orbit, in heliocentric theory. Kepler called this plane the true ecliptic, to distinguish it from the mean ecliptic, which in his physical theories was the plane of the solar equator.

Epicycle: a small circle or sphere, which rotates around its own center, while being carried as a whole around the circumference of a larger circle or sphere.

Epicyclic anomaly: the angular distance of some point on an epicycle from the aphelion (or apogee) of the epicycle. The aphelion (apogee) of an epicycle is that point on the epicycle which is farthest from the center about which the epicycle as a whole is carried. The geometric transformation which carries a simple eccentric model into an equivalent simple epicycle model also carries the eccentric anomaly in the former into the epicyclic anomaly in the latter.

Equant: an imagined circle from whose center the motion of a heavenly body appears uniform. Also, the center of an equant circle. The equant is normally used as a separate concept only if it differs from the (eccentric) path of the planet, a possibility first proposed by Ptolemy.

Equation: (Latin *aequatio*) in ancient astronomy, a small correction that was added to or subtracted from a crude approximation, typically of an angle, in order to obtain a more refined approximation. Note that this is quite unlike the modern usage of the word.

Equation of center: the correction for motion on an eccentric, rather than a concentric, circle. The equation of center is thus the difference between the mean anomaly and the true anomaly. See equation, optical equation, physical equation.

First anomaly: that irregularity in a planet's motion whose period was not related to the conjunctions and opposition of the planet. For Mars, the first anomaly was due to the variations (both real and apparent) in the motion of the planet on its own orbit. This is also called the zodiacal anomaly.

Geocentric: centered on the earth. For Aristotle, Ptolemy, and (in certain contexts)

Tycho Brake, the earth was the center of the planetary system. Lunar theory is always geocentric, of course.

Heliocentric: centered on the sun. For Copernicus and Kepler, and in certain contexts Tycho, the sun was the center of the planetary system.

In antecedentia: in the opposite direction of the circulation of the planets, disregarding the daily rotation of the heavens. From east to west; or clockwise, as seen from the north.

In consequentia: in the direction of the circulation of the planets, from west to east; counterclockwise, as seen from the north.

"Kepler motion": (A modern term, obviously) motion according to "Kepler's laws", where planets move on elliptical orbits, according to the area law.

Latitude: the angular distance of a heavenly body from the ecliptic plane. Planetary and lunar latitudes are always small.

Libration: Kepler's term (probably adopted from Copernicus) for back-and-forth motion, especially for a planet's approach to and withdrawal from the sun as it passed from aphelion to perihelion and back. The term is from libra, scales or balance, in reference to the rocking motion of a balance.

Limits: the two points of greatest north and south latitude.

Longitude: the angular distance of a heavenly body, in the ecliptic plane, from some reference point. The conventional reference point was the vernal equinox or, for Copernicus, a star near the vernal equinox.

Lunula: Kepler's term for the area, shaped like a very thin crescent moon, between an oval orbit and the circle which circumscribes it.

Mean anomaly: the angular distance a body would have from its aphelion (or apogee) if its angular motion around the sun (or earth) were uniform. Mathematically, mean anomaly is equivalent to time, with the norm that the body's periodic time equals 360°

Mean distances: the parts of an orbit midway between aphelion (apogee) and perihelion (perigee), where the distance to the central body is about equal to its mean value.

Mean sun: an imaginary body in the ecliptic whose position is where the sun would be if its apparent motion were uniform. For Copernicus, the mean sun became the center of both the earth's circular path and (since he thought the earth to move with constant velocity) the center of the earth's equant motion. Minor ambiguities arose in the use of the term after Kepler distinguished between the earth's orbital and equant centers.

Nodes: the two points where the path of a planet (other than the earth) intersects the ecliptic.

Occultation: a (rare) situation when one planet or star passes directly in front of another, hiding it.

Optical equation of center: Kepler's term for that component of the equation of center which arose from the fact that the distance between a planet and the (eccentric) center of its orbit varied, so that the apparent angular motion was slower when the distance was greater, and swifter when the distance was less. See Physical equation of center.

Opposition: the situation or moment where a heavenly body lies directly opposite the sun, and hence the earth lies between the two. In true opposition, the body lies opposite the sun itself; in mean opposition, the body lies opposite the mean sun.

Perigee: the point of least distance from the earth (normally in a geocentric context).

Perihelion: the point of least distance from the sun (in a heliocentric context).

Physical equation of center: Kepler's term for that part of the equation of center which arose from the fact that the speed of the planet's real motion actually varied. See Optical equation of center.

Second anomaly: that irregularity in a planet's motion whose period equaled the time between successive (similar) conjunctions with the sun. For Mars, the second anomaly was due to the motion of the earthbound observer. Kepler also called this the synodic inequality.

Synodic: an adjective applied to relations involving conjunctions and oppositions. A synodic month is the time between two successive conjunctions of the moon and sun, and is what one normally thinks of as a lunar month. The synodic inequality is the second anomaly of planetary motion, which is due to the earth's revolution around the sun, and which vanishes at conjunction and opposition to the sun.

Syzygies: in lunar theory, the points of opposition and conjunction with the sun. Lunar and solar eclipses can occur only at syzygy, obviously.

True anomaly: the actual angular distance of a heavenly body from aphelion (or apogee), as measured at the central body. Kepler also called this the coequated anomaly. Compare to eccentric anomaly, measured at the center of the orbit, and mean anomaly, measured at the equant center—although the former acquires different nuances in different theories.

True sun: the sun itself, when contrasted with the mean sun.

Zodiacal anomaly: the first anomaly, whose variation occurred with the planet's longitude (its position in the zodiac), after correction for the second anomaly.

Index

Index to the *Astronomia nova*

This guide shows the pages, by chapter, on which reference is made to *Astronomia nova*.